# Breeding Services for Small Dairy Farmers

## Sharing the Indian Experience

T0132675

C.T. Chacko
F. Schneider

## Science Publishers, Inc.

Enfield (NH), USA          Plymouth, UK

SCIENCE PUBLISHERS, INC.
Post Office Box 699
Enfield, New Hampshire 03748
United States of America

Internet site: *http://www.scipub.net*

*sales@scipub.net* (marketing department)
*editor@scipub.net* (editorial department)
*info@scipub.net* (for all other enquiries)

**Library of Congress Cataloging-in-Publication Data**

Chacko, C.T. (Chanthanattu Thomas), 1944-
  Breeding services for small dairy farmers: sharing the Indian experience/C.T. Chacko,
F. Schneider.
      p. cm.
  Includes bibliographical references
  ISBN 1-57808-380-X
    1. Dairy cattle--Breeding--India. 2. Dairy cattle--Genetic engineering--India. 3. Dairy
cattle--Breeding. 4. Dairy cattle--Genetic engineering I. Schneider, F. (Fritz), 1953-II.
Title.

SF196.I4C53 2005
636.2'142'0954--dc22

                                                                                    2005044056

ISBN 1-57808-380-X

© 2005, Copyright Reserved

Published by Science Publishers, Inc., Enfield, NH, USA
Printed in India.

# Foreword

Growing markets for livestock products, liberalisation of world trade, and overall global economic growth are creating unprecedented opportunities for accelerating the rate of poverty reduction in the developing countries. The structure of livestock production in a number of developing countries, including India, would suggest that the poor should be the main beneficiaries of this growth. But, at the same time, poor households with small marketable surplus face significant financial, technical, and social barriers which limit their ability to compete in the expanding markets. Capitalisation of the opportunities presented by expanding markets for the benefit of the poor therefore requires policies and programs that facilitate growth in productivity, build institutions that encourage integration of small producers in the value chain and effectively regulate markets. In the emerging market environment, the real challenge will be to provide quality products at minimum costs. It is therefore critical that adequate attention be paid to finding ways and means of reducing cost of livestock production in developing countries.

In agriculture, cost reductions are primarily driven by technology. In case of livestock this means scientific advancements in breeding practices, disease control, and overall feeding and management efficiency. In addition, enhancing competitiveness at the sectoral level requires significant improvement in public and private infrastructure in rural areas and creation of a favorable investment framework. While this is likely to enhance the competitiveness at the sectoral level, it is also likely to create more intense competition from large producers due to significant economies of scale in collection, processing and marketing of livestock products. The technological advancements must therefore take an explicit account of the production, market and institutional environment in which small producers function and make decisions. Development of such technologies can be a significant challenge in view of the complex interactions between animal genetics, human relations, social and environmental sciences, and economic and behavioral sciences.

India has a relatively long experience of designing breeding programmes for small farmers and both the authors of this book have been closely associated with the design and implementation of these programmes. I am pleased that they have chosen to systematically document their experiences in the form of this book. The book presents a very practical "hands-on" approach to designing and implementing breeding programmes for small dairy farmers. By embedding the discussion within the prevailing institutional and cultural environment of the developing countries, the authors have successfully addressed the issue of suitability of breeding technology for poor countries. In that sense, the publication represents a milestone in taking forward the literature on advancing livestock production for the benefit of the poor. I hope the practitioners and decision makers concerned with designing breeding programmes will find it useful.

**Vinod Ahuja**
*Associate Professor*
*Centre for Management in Agriculture*
*Indian Institute of Management, Ahmedabad*

# Preface

Better validation of indigenous domestic animal genetic resources is becoming more important with regard to the potential of livestock for poverty alleviation and income generation. It has been well established, that livestock resources are more equitably distributed than other production resources, such as land and capital. A large part of the rural smallholders are poor and depend on livestock for their livelihood, whereby income by selling animal products (milk, eggs, meat) is only one, though an important aspect.

To improve indigenous breeds for sustainable income and employment generation, the methods to be employed are the same as developed in systematic breeding programmes be it for crossbreeding or selective breeding within a specific breed.

The book, after an introductory chapter on global and Indian trends, systematically introduces the reader into the breed improvement theory and illustrates the theory with practical examples and case studies. The book banks on Indian experience and within India predominantly on the Kerala experience.

*"The Kerala programme with its systematic and holistic approach to livestock development, is the most successful and most sustainable livestock project I know"* Dr. Samuel Jutzi, Director Animal Production and Health Division, FAO, during his presentation on biodiversity in livestock, at the Swiss Federal Institute of Technology (ETH), Zurich, Switzerland, at the occasion of the World Food Day, October 14, 2004.

The textbook is addressed to animal science teachers, to undergraduate and postgraduate students as well as to decision makers in state and central livestock departments. The book will have a reach outside India and will be a valuable guide on animal breeding practices in any tropical country with a large proportion of its livestock within small and medium holdings. Decision makers, researcher and students will find this book a useful reference and the collection of well structured exercises on the enclosed compact disc will be instrumental to develop customised models by calculating various parameters in cattle breeding operations with case specific data.

C.T. Chacko
**Fritz Schneider**

# About the Authors

## C.T. Chacko

C.T. Chacko was the Chief Executive of Kerala Livestock Development Board, Kerala, India till 1999. After retirement he works as a consultant in livestock development issues. While working with the Kerala Livestock Development Board, he was associated with formulation, planning and implementation of the cattle-breeding policy and the implementation of the field performance recording programmes in many states in India. He has been instrumental in the evolution and development of the Sunandini breed of India. He has held many positions such as member of the advisory committee of post-graduate students in animal production and animal breeding, Kerala Agricultural University, Kerala, India, member of the subgroup to prepare the national project for cattle and buffalo breeding in India, Convenor of the expert committee to formulate the breeding policy of Kerala, India, member of the state advisory committee for livestock breeding, Uttar Pradesh, India. His publications include many research papers, which have appeared in national and international journals. He has undertaken many missions for the Swiss Agency for Development and Cooperation, India, World Bank, India, Swiss College of Agriculture Zollikofen, Switzerland, National Dairy Development Board, Anand, India and Intercooperation, New Delhi, India. He holds master's degrees in cattle production from the University of Kerala, India and in Animal Genetics and Breeding from the University of Edinburgh, United Kingdom.

## Fritz Schneider

Fritz Schneider studied Agronomy and Animal Science at the Swiss Federal Institute of Technology in Zurich, Switzerland and did post-graduate studies in Animal Science at the Department of Animal Science, University of British Columbia, Vancouver, Canada (Master of Science). He is Vice Director of the Swiss College of Agriculture (SCA), Zollikofen, and Head of the Institute SHLexpertise. SHLexpertise is the institute within SCA responsible for the development of applied research and development, services and continuing education. He is Professor for Tropical Animal Production Sciences and in this function responsible for tropical animal breeding, animal nutrition, animal husbandry and animal economics. He has spent more than seven years in India in various positions from breeding expert to chief project adviser responsible for livestock and watershed projects within Swiss Agency for

Development and Cooperation (SDC) as part of the Swiss Embassy in New Delhi. He is responsible for mandates to support development agencies for project planning, management, monitoring and evaluation (mainly livestock production, natural resource management and agricultural extension programmes) in India, Bhutan, Kyrgyzstan, Vietnam, Ukraine, Romania, etc. He continues to be senior consultant to Intercooperation and Swiss Agency for Development and Cooperation in their poverty alleviation focused programmes with major livelihood and livestock components.

*May 2005*

**C.T. Chacko**
*Kalampoor, Kerala*
*India*

**Fritz Schneider**
*Bremgarten*
*Switzerland*

# Acknowledgements

We are grateful to all those have contributed to the writing of this book. Special mention is made to:

Dr. Vinod Ahuja, Ahmedabad, India who wrote the foreword to the book, Dr. Martin Menzi, Thun, Switzerland, Dr. Fritz Schmitz, Ittigen, Switzerland and Dr. Markus Schneeberger, Langenthal, Switzerland who have reviewed the various chapters of the book and provided immense support to improve its content to the present form, Mr. Mathew C. Antony, Softscan Corporation, Ernakulam, India for the assistance in developing the necessary software to put the exercises in the right perspective, Mrs. Ana Maria Hintermann, Bern, Switzerland who did the editing and page setting before submitting the manuscript to the printers and publishers, Officers of the Kerala Livestock Development Board, Kerala, India for assisting in taking many of the pictures, M.P.G. Kurup, Bangalore, India for giving statistics on the livestock situation in India and G.S. Nair, Trivandrum, India and Ms Margaret Majithia, New Delhi, India for the English corrections.

## Photographs

The photographs in the book including the cover page were supplied by: Mr.Antonio Michael, Ernakulam, India, Mr.Theo Iff, Hinterkappelen, Switzerland, Mr. Walter Diener, Thunstetten, Switzerland. We express our sincere gratitude to them.

## Sponsors

The writing of the book would not have been possible but for the financial support received from individuals and organisations listed below. We express our deep gratitude to all, both for the financial as well as for the moral support.

### Individuals

Mr. Walter Diener, Thunstetten, Switzerland, Dr. Hans Gujer, Zurich, Switzerland, Dr. Urs Küpfer, Bern, Switzerland, Dr. Hansueli Kupferschmied, Neuchâtel, Switzerland, Dr. Martin Menzi, Thun, Switzerland, Dr. Andreas Schneider, Wil SG, Switzerland, Mr. Willi von Siebenthal, Windisch, Switzerland and Mr. Peter Wiesmann, Kirchlindach, Switzerland.

*Institutions*

BIG-X AG, Seewen, Switzerland, Helvetas, Swiss Association for International Cooperation, Zürich, Switzerland. Nestlé India Limited, Gurgaon, India, Samsoft AG, System Consulting and Development, Bremgarten, Switzerland. Softscan Corporation, Cochin, India. Swiss College of Agriculture, Zollikofen, Switzerland, Swissgenetics, Zollikofen, Switzerland, Swiss Holstein Breeder's Federation, Posieux, Switzerland and UBS, Swiss Banking Corporation, Bern, Switzerland.

*The Swiss Agency for Development and Cooperation (SDC) and Intercooperation (IC)* deserve special mention. SDC is the international development agency as part of the ministry of foreign affairs of Switzerland and as such the nodal agency for all Swiss supported livestock programmes in India. IC is a Swiss foundation specialised in international and development cooperation. IC's principal working domains are natural resource management, rural economy, local governance and civil society. IC is acting as implementing agency for most livestock oriented programmes of SDC in India since 1983. SDC and IC, via their common programme "Capitalisation of Livestock Programme Experience in India (CALPI), have agreed to make this book available to the relevant departments, educational institutions and decision makers in India.

# Contents

# List of Tables

# List of Illustrations

# List of Boxes

# List of Toolbox Exercises

# List of Annexes

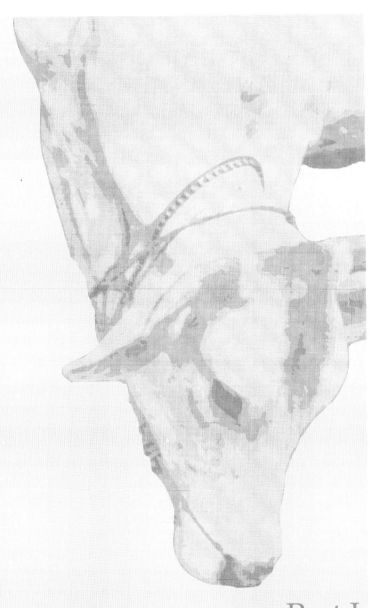

# Part I

## Approach to breeding programmes for developing countries

# 1. Trends in Livestock Production

*At the global level, a rapid increase in the demand and supply of livestock products continues to be evident. The major trends and issues related to this demand-driven development are examined. A close look is taken at India's livestock and the livestock services to set the frame for a detailed study of breeding services for small dairy farmers. Breeding systems are discussed and the merits and limitations of breed improvement within the common mixed smallholder production system in India are discussed. A special focus is given to indigenous domestic animal genetic resources, which are becoming more and more important with regard to the potential of livestock for poverty alleviation and income generation.*

# Introduction

From the early 1970s to mid-1990s consumption of meat in the developing countries increased by 70 million tons and that of milk by 105 million tons of liquid milk equivalents. The rate of increase was about twofold and three-fold respectively compared to that of developed countries. The market value of this increased consumption was more than twice that of grains such as wheat, rice and maize. The past trend in the growth of population, urbanisation and income that fuelled this increase in the consumption of these livestock products is expected to continue well into the twenty-first century, creating a veritable Livestock Revolution (Delgado, et al., 1999, Delgado, et al., 2001). As these events unfold, farm income could rise dramatically, but whether smallholders and landless animal keepers will share that gain is still uncertain and will depend largely on the availability and quality of services they can receive or organise for themselves.

# Major issues

## Demand-driven increase of livestock production

The livestock revolution is propelled by demand. Though consumption of livestock products in developing countries is increasing from the very low levels of the past, it still has a long way to go before coming near developed countries' averages. Per capita growth is faster in regions where urbanisation and rapid income growth result in people adding variety to their diet.

## Rapid increase of industrialised livestock production units

The rapid rise in livestock production in developing countries has been confronted in recent years by dwindling grazing resources, a shortage of roughage for ruminants and a fast-growing demand largely centred on rapidly growing megacities fuelled by the development of sectors other than agriculture. This increases pressures for industrial approaches to satisfy the urban demand. Together, these trends help explain the large share of pigs and poultry in the production increases in both North and South. These industrial units can be characterised as follows:

- Land detached (feed, waste), mostly urban or periurban
- Capital intensive commercialised and large-scale units

*India has the largest number of dairy co-operative
societies in the world: A milk collection point of a
dairy co-operative society in India.*

- Highly specialised, single purpose, meat, egg or milk production.
- Efficiency in resource utilisation, specialisation and uniformity fosters efficiency in terms of input units per output unit
- Often highly polluting due to nutrient surpluses and gaseous emissions
- With regard to animal genetic resources, industrial systems use uniform genetic stock favouring genetic erosion both within breed and within species. (Loss of genetic variation, fitness traits and disease resistance.)

## Marginalisation of small and landless livestock keepers

The rapid increase in demand for livestock products poses challenges to the traditional production systems. Rural smallholders are likely to be marginalised, as they will not be able to ensure steady supply, either in quantity or quality. Their market access will not therefore increase automatically with increasing demand. Further, the necessary technology shifts are expensive and access to the required credit is often poor for smallholders. These are some of the reasons why there is an imminent danger that small producers become supplanted and marginalised and income disparities may grow. On the positive side, there is some employment generated in industrial production and substantially more employment is generated in the value-added food chain. Smallholders need to have more options to improve competitiveness if they wish to compete in the domestic and more so international markets. Technology can play a role in this, as can understanding of the determinants of economies of scale and the impact of biotechnology on structural change. Enforcement of environmental, public, and veterinary health regulations also affects smallholder competitiveness and needs to become more transparent for the smallholder communities (Hall, 2004).

For labour intensive commodities (e.g. milk production) smallholder livestock production in Asia has a cost advantage over large-scale industrial systems due to the availability of low-cost family labour and relatively modest economies of scale in livestock production. But, high feed deficits, scarcity of land, high cost of capital and high delivery cost of specialised inputs such as vaccines and drugs tend to erode the competitive advantage conferred by low labour costs. Further, there are significant problems with product quality, safety and uniformity, compounded by widespread prevalence of trade-preventing and production-limiting diseases. Poor infrastructure and poor access to services and disabling government regulations on producer organisations impose further transaction costs on the smallholder producers, while processors and other economic agents undermine the regions ability to compete in growing markets for livestock products. Thus, the key constraint faced by smallholders is restricted market access which, in turn, results in lower prices for their products (Ahuja, 2004).

In the light of these factors, policy measures to improve sector competitiveness will need to focus on

- Improving public and private infrastructure in rural areas
- Effective disease control
- Enhanced research and development
- Creation of a favourable investment framework
- Awareness creation and education about product quality
- Refinement and implementation of sanitary and quality standards that can be met by smallholders.
- Capacity building of small holders to enable them to meet quality standards

Looking at WTO the trade distortions for agricultural and animal products are substantial and in general are negative for the smallholder producers in the developing world for reasons described above. The ongoing negotiations will need to focus both on international trade policy as well as on major domestic policy reforms. An excellent overview of these issues can be found in *Livestock and Livelihoods. Challenges and Opportunities for Asia in the Emerging Market Environment* (Ahuja, 2004).

## Livestock and environment interactions

While the industrialised livestock production units are in a position to cope with the increasing demand, they often pollute the environment if not properly managed. As animal production systems intensify, there is a potential for negative environmental impact from improper storage and application of manure polluting surface and groundwater. The problem results from improper management rather than from the animals per se, because the same problem, for example, can occur with improper use of chemical fertiliser in intensive agriculture. In addition, there are other threats such as:

- Nutrient depletion in feed-producing areas
- Severe exploitation of non-renewable resources
- Erosion of animal bio-diversity
- New and newly emerging animal diseases
- New disease threats for humans
- Food contamination with microbes, medicaments and toxic residues

In most countries, there is no effective legislation to check these hazards. The multinational initiative 'Livestock Environment and Development' (LEAD, 2004) studied these aspects and recommends the development of systems based on the "polluter pays—provider gets" principle which would increase the cost of industrial production and in turn render the non-polluting smallholders more competitive.

## Livestock services for small livestock keepers

About 900 million of the world's 1200 million extremely poor people live in rural areas. Most of them rely on agricultural activities for their livelihood. Livestock keeping is crucial for the rural poor. Approximately 600 million poor smallholders keep nearly one billion heads of livestock. An IFAD publication (2004) examines livestock services for the poor, identifies the major issues and recommends an appropriate course of action (Table I.1).

In addition to these issues, smallholder livestock producers often face serious marketing constraints. Apart from inadequate access to services they also lack proper access to markets and timely market information. Due to resource constraints, seasonal and climatic conditions the smallholders often are not in a position to maintain a steady level of production with regard to quantity as well as quality. The emerging markets, however, demand a steady flow of quality products. Such constraints can only be overcome by forming farmer based organisations (co-operatives, associations, etc) or by efficient brokers who are accountable to the smallholders.

**Table I.I**  Livestock services for the poor. Crosscutting issues and recommendations

| Issues | Recommendations |
|---|---|
| **Public and private sector roles.**<br><br>Many service delivery systems in developing countries are undertaken by a public sector that is involved in too many tasks. The public sector does not reach the poor and, at the same time, hinders the emergence of more efficient private sector delivery because it competes and overlaps with private sector operators. | **Introduce policy debates on perceptions** of the roles of the public and private sectors in service delivery for the poor and develop public-private partnerships, whereby the public sector delegates decision-making on the scope and content of the delivery of services as a public good. This would include full or partial funding of the services by the sector, but preferably through subcontracting to private operators and the introduction of voucher systems, or other competitive grant system for research and education. |
| **Gender imbalances in livestock services.**<br><br>Conventional livestock services are provided by men for men, whereas livestock production often plays a crucial role in the livelihood of poor women. If poverty is to be reduced, women must become involved in livestock services both as producers and service providers. | **Targeted involvement of women** both as producers and service providers so that their capabilities and social status can be enhanced. |
| **Inappropriate technologies are employed in livestock services.**<br><br>The provision of appropriate technology also requires more coordinated effort. The typical areas of importance to the poor (small livestock, improvement of local breeds, thermoneutral and long-lasting immunity-producing vaccines, low-labour input systems for fodder production) are neglected. Experience until now indicates that people adopt technology if they possess adequate resources, labour and skills and if the technologies are beneficial to their livelihoods. | **Involve poor livestock keepers in technology generation and transfer** so that they can gain more experience in pro-poor livestock technologies. The necessity of building on the knowledge, skills and resources that poor livestock keepers already possess must be emphasised. |
| **Knowledge and learning systems are absent within livestock services.**<br><br>Livestock services have typically been based on conventional discipline-specific knowledge biased in favour of animal health and sometimes genetics, but do not supply opportunities for learning, nor a holistic approach towards smallholder livestock systems. | **Develop knowledge and learning systems within livestock services** that can strengthen the capacities of livestock keepers to demand or seek information, training and advice. In order to implement the concept, the building up of professional capacities in livestock advisory services must be emphasised. |
| **Lack of access to financial services.**<br><br>In most areas, poor livestock keepers lack access to financial services. Even where such services are provided, the impact of credit schemes in reducing the poverty of the poorest is not usually considered. The poor still face barriers in benefiting from the services. | **Research on financial services** that are appropriate to the poor should be carried out. The impact of credit and debt on the poorest livestock keepers and ways to help them benefit from microfinance systems need to be more clearly understood. |

*Sources: Livestock Services for the Poor, IFAD, 2004.*

# India's livestock sector

Liberalization and a fast-growing economy have led to changes in the livestock and dairy sectors in India. Several, sometimes contradictory, trends emerge. The importance of local knowledge systems in the livestock sector, as well as the recognition of the relevance of local breeds, is slowly gaining ground in a discussion on livestock. It remains to be seen, however, how far these trends will be able to counterbalance market- and demand-driven developments that clearly favour high-yielding, pure exotic animals or their crosses. The private sector is already strongly involved in poultry and also in feed manufacturing, the leather industry and, lately, also in dairying, as demand for livestock and dairy products is growing faster than the population. The private sector, while responding to this demand, makes investments largely concentrating on irrigated rural as well as periurban areas with high potential for livestock production (milk, meat and eggs) to the detriment of less-endowed marginal areas. In recent years, there has been an increase in intensive, high-input dairy farms operating close to urban markets that are supported by a corresponding increase in concentrate feed production, creating environmental problems. These farms demand genetic material with high production potential. The major issue that will affect the future of livestock production systems in India is, therefore, the extent to which the role of government and semi-government organizations in service and marketing will be reduced and the sector opened for private interventions. While the private sector can be expected to improve efficiency and productivity, the impact of liberalization on the weaker sections of the population as well as the environment may be less positive. The gradually evolving picture of increasingly liberalized and high productive dairy and livestock sectors, therefore, calls for structured approaches in livestock services delivery systems.

Livestock production systems in India today are characterized by widely distributed smallholdings with almost 75% of rural households owning livestock of one type or the other. Livestock is closely integrated in the crop husbandry components of the traditional farming systems and both are complementary to each other. In the low capital input but labour- intensive production systems, livestock provides—in addition to major products such as milk and meat—equally important outputs like manure and draught, while crop residues provide feed and fodder for the animals. The livestock sector also includes a range of non-farming activities such as procurement, processing, transporting and marketing of livestock and livestock products and provides employment opportunities to millions (SDC/IC, 1995b).

Numerically, the livestock wealth of India is impressive and represents a sizeable part of the world livestock population. For the main species, the 1956,

*Selling milk at the doorstep is a system in the villages of India*

1998 and 2003 figures are given in Table I.2. The most numerous livestock in India are cattle, comprising 190 million animals, an increase by 25% of the population since the 1950s. In the same period, the number of buffaloes in India more than doubled to over 96 million—well over half the buffaloes in the world. The growth in goat population has been even greater, from 55 million in 1956 to 120 million in 2003, while sheep numbers rose from less than 40 million to almost 62 million heads. Since independence, the contribution of agriculture to the Indian GDP has steadily declined from 56% to 20.5% in 2004. The contribution of the livestock sector to agriculture, on the other hand, has consistently increased since the early seventies to the present level of around 25%, or just below 10% of GDP (AH Series-9, 2004). Cattle and buffaloes clearly dominate the economics of the sector and milk accounts for 67% of the total value of livestock products in the GDP. In absolute figures, the increase in milk production since independence has been impressive, increasing from 17.4 million tonnes in the early 1950s to more than 91 million tonnes in 2003. But the overall average productivity of animals has remained comparatively low. India produces 14% of the world dairy output despite its 19% share of the world's bovine population (AHS Series-9, 2004).

**Table I.2**  Livestock wealth of India (in millions)

| Species | India 1956 | India 1997 | Change 1956 to 1997 | India 2003 | Change 1997 to 2003 | World 1997 | World 2003 | India as % of World in 1997 | India as % of World in 2003 |
|---------|-----------|-----------|---------------------|-----------|---------------------|-----------|-----------|-----------------------------|-----------------------------|
| Cattle  | 158.7 | 198.9 | + 25 %  | 187.4 | - 1.18 % | 1308.2 | 1331.5 | 15.2 % | 14.1 % |
| Buffalo | 44.9  | 89.9  | + 107 % | 96.6  | + 1.45 % | 159.1  | 170.4  | 57 %   | 57 %   |
| Goat    | 55.4  | 122.7 | + 118 % | 120.1 | - 0.43 % | 681.6  | 765.3  | 18 %   | 16 %   |
| Sheep   | 39.3  | 57.5  | + 44 %  | 61.8  | + 1.45 % | 1043.8 | 1024.7 | 5 %    | 6 %    |

*Sources: Govt. of India; AH series 9, 2004; FAO, 2005*

## Distribution of livestock

Livestock holdings are more equitably distributed than land resources in India. Sixty per cent of livestock owners are marginal and landless farmers who cultivate less than 12% of the total land under production. For example in Kerala, most of these small farmers keep exclusively female animals for milk production and derive a major part of their cash income from milk sales (Wälty, 1999). These facts support the claim that livestock-related interventions can indeed be a successful strategy for poverty alleviation, provided the issue of participation by the poorer segments is duly ensured.

## Livestock in the livelihood system of smallholders

Smallholders keep livestock for various reasons, e.g.:

- Agricultural production (to provide draught and manure)
- Generation of regular income (through sale of products such as milk)
- Risk avoidance (as assets for emergencies)
- Cultural factors (status, religious importance)

While large ruminants have always received considerable government attention and funds, interventions in small ruminant breeding or management are limited both financially and in terms of innovative actions (Kurup, 1995). Overall bovine population dynamics show a clear trend towards more adult females in relation to working males (NLP, 1996). This change correlates closely with the spread of mechanization in Indian agriculture and the growing importance of milk production as an income-generating activity for rural households. The latter is also the main reason for the increase in numbers of buffaloes of which 87% are females. Regional differences, however, are considerable. While the foregoing holds true for the intensive agricultural zones in the northern and southern states, the extensive production systems of the Central Deccan Plateau and the eastern states still largely rely on animal draught and, consequently, a high ratio of male animals. Next to crop production, animal husbandry is the most important income-generating activity in agriculture. Livestock, therefore, is an important means to improve the livelihood of smallholders (Lehmann, et al., 1994; Adams, 1996). In spite of concerted efforts of government institutions, non-governmental organizations (NGOs) and foreign collaborators to improve access to breeding services, still only 13% (24.6 million) of the cattle population and an even lower number of buffalo, goat and sheep are 'improved' as a result of using quality breeding material in the form of progeny-tested frozen semen or selected male animals. The major reason for this low coverage is lack of access of the smallholder to the required supplies and services. Nevertheless, in 2000, the share of cattle milk produced by crossbreds has been 41% showing a significantly higher production of the improved animals (AHS Series-9, 2004).

An estimated 73% of all rural households in India—60 million households own cattle, buffalo or both (NLP, 1996)—depend on livestock as a major source of supplementary income. The livestock sector has been one of the few sectors projecting growth in India over the past five decades, milk being the major product, both in terms of quantity and value, among all farm produce. This steep increase was brought about by a rise in income countrywide, the market stimulus created by an expanding milk cooperative network in villages and the production inputs provided under government and cooperative programmes. During the same period, smallholder milk production also went through a gradual transition from primarily satisfying home consumption needs to surplus milk being available for sale in the market.

Smallholder production will continue to develop with increasing demand for dairy products in the urban sector. Despite the sizeable contribution of the country's 190 million cattle and 96 million buffaloes to the total income from the livestock sector, the national average productivity per animal is low (700 kg of milk per lactation for cattle and 1550 kg for buffaloes), mainly due to the poor genetic quality of animals and feed inadequacy of resources (AHS Series-9, 2004). Smallholdings, averaging 2-3 animals, are found throughout the country. Indiscriminate breeding, use of unproven sires for artificial insemination, lack of culling and laws prohibiting slaughter of cattle in most of the states, have adversely affected genetic progress. Intensive bovine production is generally concentrated in the higher rainfall regions, areas characterized by high population densities. Increased human population has decreased the amount of land available for crop production per holding, thereby reducing the availability of crop residues and by-products as sources of animal feed. The growth in livestock numbers, coupled with shrinking traditional grazing areas has put intense pressure on the existing grazing areas, encouraged encroachment on forest lands and ultimately contributed to the degradation of land resources (World Bank, 1996). India, like many other developing countries, now faces an immense task of designing key strategies and implementing appropriate policies to achieve a sustainable balance between livestock production and environmental conservation (Kurup, 1998a; Steinfeld et al., 1997). In order to reduce the number of bovines and still be able to meet the country's rapidly growing demand for dairy products, the genetic potential of animals has to be improved.

## Land and livestock holding in India

Land holdings in India are in general small and fragmented; medium and large holdings account for less than 10% of the holdings. Table I. 3 shows that three-quarters of the ruminant livestock of India are found in holdings of less than four hectares. Three-quarters of the holdings are smaller than two hectares. The result is a rather high animal density within these smallholdings. The table further confirms that production intensification in livestock needs to focus on smallholdings which already have a high intensity of animals per holding and animals per hectare. Distribution of land is grossly inequitable: marginal and small holders account for over 78% of the holdings, but they own/operate less than 33% of the total farming land. With the relentless growth in human population, the number of holdings steadily increases and consequently the size of land holding in general has been shrinking over the years, making individual holdings progressively unviable. In 1961, the average holding size for all categories together was 2.52 ha per holding, but by 1992 this figure has shrunk to 1.34 ha average per holding. Diversification in agriculture thus became an unavoidable compulsion for the vast majority of the farming community in order to protect livelihoods. Livestock invariably was their first option for diversification.

As mentioned above, livestock holding in general and milch animal holding in particular, appear to be far less iniquitous compared to land holding: marginal

**Table I.3**   Ruminant livestock distribution per holding in India

| Category | < 1 ha | 1 - 1.9 ha | 2 - 3.9 ha | 4 - 10 ha | >10 ha | All Holdings |
|---|---|---|---|---|---|---|
| Ruminant animal population % | 37.30 | 19.10 | 21.30 | 17.00 | 5.30 | 100.00 |
| Average size of holding (ha) | 0.40 | 1.45 | 2.76 | 5.90 | 14.30 | 1.66 |
| Animals per ha | 5.80 | 2.60 | 1.90 | 1.20 | 0.60 | 2.20 |
| Animals per holding | 2.33 | 3.70 | 5.47 | 7.29 | 9.16 | 3.70 |
| Distribution of holdings (%) | 57.20 | 18.40 | 13.90 | 8.30 | 2.10 | 100.00 |

Source: World Bank, 1996.

and smallholders together owned over 67% of all milch animals in 1992. The *Gini Coefficient*[1] representing the index of inequity in ownership of dairy stock shows a perceptible decline from 0.43 in 1961 to 0.37 in 1971 and further to 0.28 in 1991. Milch animals among crossbred cattle also tend to concentrate (78 %) in the marginal and smallholdings (Fig. I.1).

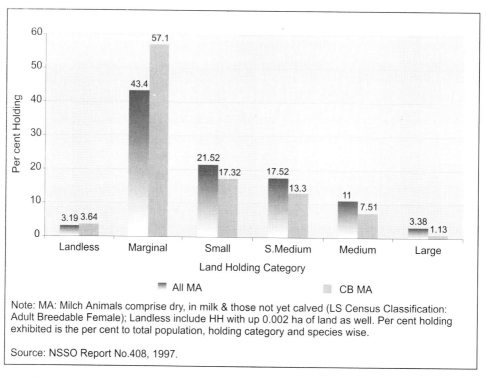

**Figure I.1**    Distribution of milch animals in rural households, landholding category wise: 1992

Large ruminant holding per household varies considerably with the regions both in number and species held (cattle mostly in the south-east, buffalo in the west and cattle as well as buffalo in the north), but the average holding seldom exceeds three to four animals per household. Large ruminant holding size as a rule is greater in Punjab, parts of Haryana and western Uttar Pradesh (trans-gangetic plains). The few large dairy farms in the country are mostly institutional farms or commercial dairy farms for milk production in metro cities and other major urban agglomerations. In Gujarat, Rajasthan, Punjab, Haryana, Himachal Pradesh and Jammu and Kashmir, there are migrating tribes

---

1. The Gini coefficient, invented by the Italian statistitian Corado Gini, is a number between zero and one that measures the degree of inequality in the distribution of income in a given society. The coefficient would register zero (0.0 = minimum inequality) for a society in which each member received exactly the same income and it would register a coefficient of one (1.0 = maximum inequality) if one member got all the income and the rest got nothing.

(Bharwad, Rabari, Gujjar, etc.) who own large herds of cattle and buffalo, often numbering hundreds of heads. The numbers of these transhumant tribes are, however, small and fast diminishing, as grazing lands, common or private, are steadily vanishing or becoming inaccessible.

Distribution of small ruminant, pig and desi poultry follow more or less the same pattern as in the case of bovine: 86.6% of sheep and goat, over 90% of pig and desi poultry are owned by the marginal/smallholders and the landless, with marginal holders alone accounting for nearly 75% of the desi poultry. Goats are all over India and are usually held in small flocks of 2 to 4, except in migratory flocks owned by transhumant graziers with flock size in hundreds. Sheep are localized in specific areas in the country and are held in somewhat larger flocks of 10 to 20 heads and among the migratory graziers in large flocks of upto a thousand heads (tribes herding goat and sheep are: Gaddis and Bakrewals in north India and Bharwads and Rabaris in Gujarat and Rajasthan). Pigs in general are farmed traditionally by the socially backward communities under free-ranging conditions, except in the northeastern states, Goa and Kerala, where pigs are part of the mainstream farming activity and are farmed by all communities as an economic occupation, in the backyard system. Holding

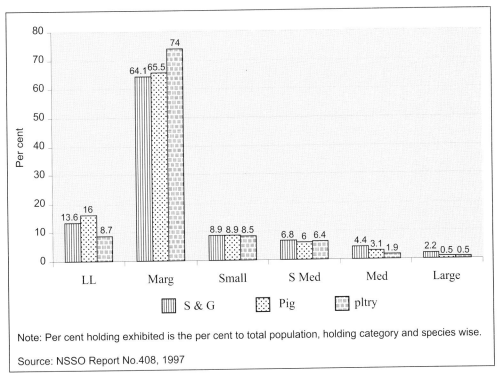

**Figure I.2**   Distribution of sheep, goats, pigs and desi poultry, landholding category wise: 1992

size of pigs per household under the range conditions is one or two and in backyard holdings 3 to 5. Desi poultry, fowls as well as ducks, are all in the backyard units scattered across the country and vary in flock size from 10 to 30. Highyielding commercial hybrid fowls account for only 35% of the total fowls and are exclusively farmed in the commercial stream of the highly efficient and modern Indian poultry industry.

Livestock has its share in the increasing ecological problems. Growing numbers of animals have led to a very high density of 2.9 units per hectare of the net sown area. Lack of adequate fodder resources in ecologically fragile areas leads to overgrazing which results in the degradation of large areas of land (SDC/IC, 1995a; Schneider, 1999). The ban on cow slaughter in force in most states is often cited as the main cause for the enormous cattle population. There is, however, no empirical link between the ban on cow slaughter and cattle numbers in different states. The ban, nevertheless, has its effects. It constitutes, for one, a substantial burden on feed and fodder resources and the full economic potential of cattle cannot be realized since animals that are neither fit for draught nor for milk production cannot be legally sold for slaughter (SDC/IC, 1995a). Further, this affects production economics of farmers as cash returns are only from milk and draught; meat has no value.

## Livestock and gender

Women play a significant role in providing family labour input for livestock keeping. In poorer families, especially, their contribution often exceeds that of men (George and Nair, 1990). However, traditionally women have a weak position in relation to decision-making regarding the utilization of income from livestock. In addition, the service and input delivery system is male dominated, which makes most of these services difficult for women to access (Kurup, 1995).

## Public livestock services

The Central and State departments of animal husbandry (and veterinary services or dairying, depending on the State) dominate support structures in the livestock sector. The departments continue to be the main actors in the field of input and service delivery based on the traditional transfer of technology concepts. Most of the services are free of cost and inputs are usually subsidized. A major part of the departmental budgets is, however, establishment costs and infrastructure, leaving little for operational expenses. Consequently, new forms of service delivery systems are being developed and tested in various states of India (Morrenhof et al., 2004).

**Box 1. 1**   Trends in Indian livestock production

**Growing demand for livestock products.** In a fast-growing economy, more and more people have extra money to spend. The demand for livestock products is growing faster than the human population.

**Growing milk market.** Annual increase in milk market by 5 - 10% (2003: 55% buffalo milk, 42% cattle milk, 3% goat milk).

**Specialization in dairying.** Specialized dairy farms are being set up, mainly in periurban areas and in areas where crop residues and industrial by-products are abundant. These farms demand genetic material with a high production potential.

**Livestock population.** Female buffalo population is increasing by 2.5% and crossbred cow population by 5% per year. There is a trend to keep large ruminants more for production than as an asset. There has also been a decline in adult males, from 55 to 45% and an increase in small ruminants because of higher demand for meat (NLP, 1996). An increasing trend towards keeping of milch animals (graded buffaloes and crossbred cattle) by landless people has been noticed in the irrigated areas of Andhra Pradesh (Lehmann et al., 1994). In Kerala, the total number of cattle has come down from 3.4 million in 1996 to estimated 2.25 million in 2003 (AHS Series-9, 2003) and the males are only 8.6% of the population (Government of Kerala, 2003).

**Ecological awareness.** Awareness of the need to address ecological consequences of intensified livestock production is growing. Efforts are being made to link interventions in livestock and dairy sectors with improved land use.

**Conservation of livestock biodiversity.** Awareness of the need to conserve livestock biodiversity is growing among government agencies, foreign collaborators and partly with NGOs. The need for the animal owner to derive an economic benefit from such conservation measures is widely acknowledged, but specific suggestions on how to achieve this goal are yet to be developed.

**Industrial livestock production in periurban areas.** Industrial livestock production is growing fast in periurban areas of the metropolitan and large cities of India, poultry being the major sector, followed by intensive dairy operations.

# Potential of breed improvement

Breeding interventions cannot be looked at in isolation from other interventions such as management, health, and nutrition. In evaluations of livestock projects, however, the demand to look at the specific impact of breeding interventions on various socioeconomic parameters is consistently raised and is still rising. The increase in genetic production potential through breeding interventions is, compared to other interventions in livestock production (health, nutrition, management), cumulative and in that sense more sustainable than other interventions. Moreover specific studies (Weller, 1999; Nitter, 1999) show that on the national or macroeconomic level investments in breeding interventions are highly profitable, mainly because the progress made is additive and is transferred to subsequent generations without substantial additional investments or genetic maintenance costs. On a microeconomic level (animal level or smallholder level) the additive character of breeding progress has to be viewed mainly as a production potential manifested only if combined with related inputs in nutrition, health and management. The impact of crossbreeding cattle and upgrading buffaloes—either through natural service or artificial insemination—is characterised by wide regional variations. Positive examples are states like Kerala, where more than three-fourths of the animals are crossbreds. Nationwide, however, the number of crossbred adult female bovines today is approximately 24 million and thus far has not induced a noticeable trend of reducing local cow populations (AHS Series–9, 2004).

Efficiency of production is the objective of improvement. In economic terms this means that any increase in output, whatever the product, should be related to the cost of the inputs. Improvement will have taken place only if the value of the output exceeds the costs of the inputs. This can be consistent with increasing the production of milk, meat, wool, or whatever, but it is not exactly the same. There are other criteria by which the benefits of livestock production and of improvement schemes can be judged. A national goal might be to optimise production from the available resources of land, feed, or labour. Emphasis will have to be placed on sustainability of production. Whatever the aim, outputs should not be divorced from consideration of such inputs as land, feed, labour, capital, veterinary services, breeding and other items required for animal production. It is not always easy to estimate the cost of these inputs accurately, but it is important to recognise that they exist. This may appear self-evident when the inputs have to be directly paid for, but it is equally true when, for example, grazing on common property resources is practised. Even in that situation economic improvement can be said to have occurred only if a given area of land produces more animals or animal products than it did before and in a sustainable manner. The returns obtained from animals (milk, meat, work, wool, hides, dung and other products) are easier to quantify (Schneider et al., 1999).

# Limitations to breed improvement

In the past, livestock improvement—and genetic improvement in particular—have aimed simply at producing high-yielding animals. Often, high-yielding animals are more efficient in economic terms but not always, especially in situations in which feed and other resources are scarce, climate is potentially stressful for animals and the cost of securing high individual yields may be too high compared to the returns. Where flocks and herds are large, or when considering a large area (perhaps a whole country or region), it is improvement in production from the whole system in relation to the inputs required—that is more important than measurement of the increase in yield of the best individuals. The overall objective of increasing the efficiency of the animal production system will often have to be attained by bestowing attention on some particular components. However, the first step is to decide what is meant by 'improvement' in any particular situation. The reasons why several improvement programmes have failed can be summarised as follows:

- Unclear breeding aims
- Unrealistic and too complex breeding aims
- Breeding for traits which are difficult to quantify and to record
- Changing the breeding aims too often
- Breeding without due consideration to the real needs of the livestock keepers
- Overestimation of the importance of breeding in livestock systems

In many developing countries, application of modern animal breeding techniques has to overcome a number of special problems which make it difficult to achieve tangible progress. The special difficulties arise mainly from natural and economic constraints. It is important to bear these in mind to avoid unrealistic expectations from breeding programmes.

## Environmental constraints

### Feed

Typically in the tropics, the feed base, in particular that available for ruminant production, is poor or very variable due to seasonal variations in feed supply during dry and wet times of the year. Periods of prolonged drought or other disasters can drastically reduce the supply of feed from natural grazing or other crops. The cost of fertilisers and pesticides may preclude their use and further reduce the availability of crop by-products.

### Climate

Climatic extremes, hot or cold, have stressful effects on animals and hence on their performance. Many indigenous breeds are relatively well adapted to such conditions but breeds native to other environments are often not well adapted (genotype-environment interactions).

### Health

A high incidence of animal disease and high mortality rate are common constraints in animal production. Disease and mortality often occur or increase due to inadequate veterinary support and are made worse by poor nutrition. Animal diseases also restrict other opportunities for genetic improvement as most countries impose legal restrictions on the import of breeding stock—unless free from specified diseases. These regulations are intended to stop the spread of some dreadful animal diseases, but they also have the effect of restricting the use of imported superior stock for breed improvement and cross-breeding. Although the regulations apply to the import of stock from all countries, they make it more difficult for tropical countries, where many of these diseases are endemic. The use of frozen material, in particular frozen embryos, may ease the restrictions on importation of stock in future. Apart from direct effects on productivity, poor animal health has adverse effects on the rate of any genetic progress to be made in a breeding programme by:

- Delaying the onset of breeding
- Lowering the reproductive rate
- Increasing mortality

## Structural constraints

### Legal and political frame conditions

Animal breeding programmes, on a national or regional scale, depend for their success on the ability to multiply and transmit genetic improvement across the largest possible number of flocks or herds. Factors creating obstacles to the implementation and cost effectiveness of such schemes are:

- Sanitary and political restrictions on transport of animals or semen.
- Barriers to successful implementation of artificial insemination programmes (involving the production, storage, distribution and effective use of semen);
- Absence of adequate market information.
- Lack of adequate infrastructure.
- Lack of defined livestock policies and improvement strategies

### Herd/flock size

In many countries, including tropical countries, numerous small herds or flocks exist. Smallholder farmers with only one or two cows, or with a very small number of sheep, goats or pigs are common. When the number of animals involved is large, progress in animal breeding is easier to achieve than when numbers are small. Also, the improvements wanted are likely to occur in a reliable way only if the numbers involved in the breeding programme are adequate. It is more difficult to extend the full advantages of genetic improvement to small units. For breeding purposes it is therefore useful to associate small herds or flocks with larger groups.

### Animal resources

Absolute yields from many of the zebu breeds of livestock are often low—though an exotic breed may be no better or perhaps even worse under those specific conditions. Just as it is wrong to expect too much from exotic breeds when they are transposed to unfavourable conditions, it is wrong to dismiss the genetic potential of local breeds simply as 'low' and regard it as a major constraint for improvement. The concept of genetic potential should always be related to the conditions under which the animals are expected to perform. Wrong expectations interfere with the proper planning of improvement schemes and lead to disappointment over the results.

### Human resources

A shortage of adequately trained specialists creates problems in the design and implementation of breed improvement schemes. It is important to match any scheme to the availability of competent people to implement it. Capacity building, therefore, is a key element in livestock improvement programmes.

### Record-keeping

Recording and monitoring of the performance of livestock is one of the prerequisites for livestock improvement. Absence of records is a barrier to the implementation and evaluation of breeding schemes.

# Methods for production improvement

There are many ways of changing the productivity of livestock, whether judged on the basis of efficiency or yield. These include: feeding, management (including the physical environment), health care, physiological or pharmacological intervention and animal breeding (genetic improvement).

As a first step it is always best to consider the resources available for animal production and their limits, then match them against the animal breeding objectives. Later, improvements in any or all the resources (for example feed or management capability) can be matched by further improvements in the genetic capabilities of the animals. To start the process the other way round—to try to match the resources to the assumed genetic potential of the animals (for example an imported exotic breed or its crosses)—is risky and more prone to failure. The extra resources required for sustaining the improved genetic type, for example the new breed or cross, may not be consistently available and the costs may be too high.

Genetic improvement cannot be achieved free of cost either, but once attained it can maintained without further efforts. Also, each step of genetic improvement can be added to the one that went before. The benefits from genetic selection add up over time; they are cumulative. Most other forms of improvement require the whole of the input to be made each time the improvement is wanted, for example the supply of extra feed or veterinary medicines. Genetic change can be achieved by:

- Substituting one breed for another
- Crossbreeding
- Inbreeding
- Selection (within a breed or a population, e.g. a herd)
- Gene transfer (not yet at the stage of regular application to animal breeding in unfavourable environments) and
- A combination of any of these.

From among the main options listed, genetic selection (within breed and within crossbred populations) is the only practical way of creating something new as distinct from something already in existence somewhere. Genetic selection moves forward in small, cumulative steps, and thereby allows any change in feeding and management to be assimilated slowly and steadily. In the long term, selection may provide the most secure option for sustainable improvement. Unfortunately, selection, particularly in the tropics, is not often given serious consideration. This is because its immediate effect is rarely as dramatic as, for example, changes through crossbreeding. However, systematic selection programmes are very demanding and require a high level of skills, dedication and persistence from the organisations and persons involved in their implementation.

Heredity and environment may interact. Two or more breeds that do not differ much in one set of conditions may differ quite significantly in another set of conditions. People often point to the very high milk yield of Holstein cows under the best conditions in North America and contrast it with the low yield of a zebu breed or any other local breed in India. Such a difference in yield is due only in part to the difference in breed and is largely due to the major differences in the environment (climate, feeding, management practices, disease incidence).

The relative importance of breed and environment in influencing milk yield cannot be assessed if each breed is kept only in its own traditional set of conditions. When that happens, the genotypes (breeds) and the environment in which the animals are kept are said to be confounded. Likewise, if one breed is kept in one herd and the other breed in another differently managed herd, breed differences get mixed up (confounded) with other non-genetic differences between the herds.

# Breed conservation and development

During the early periods of domestication animals moved from one place to another along with their masters and isolated populations became specifically adapted to their environments. Controlled mating and selection of preferred genotypes followed in defined populations. During the 1950s artificial insemination with its advantages for application of new selection methods based on quantitative genetics has contributed significantly to the fast production enhancement in developed countries. In the race for maximum exploitation of the benefits of the specialised breeds, many local well-adapted breeds were less cared for.

Breeding of indigenous cattle and buffaloes is the traditional subsistence strategy in many mixed farming systems with a majority of small farmers, in principle, well attuned to their natural resource base. It is an important step forward to alleviate poverty among the rural population with limited or no land resources. The principal goal of modern conservation biology as expressed in the world conservation strategy (ICUN, 1980) is the maintenance of genetic variability achieved through proper population management. Poor population management, e.g. skewed selection pressures and inbreeding, results in reduction of genetic variability leading to a decline in adaptability and fitness. Genetic resource management is the priority for developing countries, as they seek to find their optimum livestock from among the native and imported genetic material (Hodges, 1986). Land (1986), however, does not support the genetic insurance value of breed conservation on the ground that even seemingly uniform populations of single breeds retain plenty of variability through normal mutation rates.

There is no generally accepted definition of a breed and there are not many breeder associations for indigenous tropical breeds. In the tropics many local cattle populations vary widely in their general appearance although they are of the same type and origin. A group of animals that, through selection and breeding, have come to resemble one another and pass those traits uniformly to their offspring are classically referred to as a 'breed'. The term breed has come from the common usage by breeders for their own purpose and is a consensus term. The definition of any breed for that matter is the one given by the respective breeders. The concept of a breed has been changing in the second half of the twentieth century from uniformity in phenotypic characters and conformation to excellence in economic qualities. For breeder organisations the breed has to provide maximum output at minimal cost. To achieve this the breed has to evolve as one suitable to the environment and produce optimally with minimum inputs.

A breed is declared endangered when its population size is below the accepted level. The minimum number of animals required for cattle to survive as a breed without risk of extinction in the near future is estimated at 1000 adult females (Maijala, 1982). However this number is much higher in developing countries because of the risks in population loss due to unexpected causes. FAO uses the working rule of studying the population for survival risks when the population size is shrinking towards 5000 adult females (Hodges, 1990).

Almost all of the breeds, strains and types of livestock of all species in India still exist in their respective breeding tracts, though the population size varies for individual breeds and many of them are on the decline with a few facing the threat of imminent extinction as well (double humped bactrian camel in Ladakh, Punganur cattle in Andhra Pradesh, for example). Centuries of neglect, indiscriminate breeding, malnutrition, ravages of epidemics, shrinking habitat and, in some species, redundancy collectively undermined their productivity as well as social and economic relevance.

The evolution of all of these breeds among all species—though initiated by human endeavour in the distant prehistoric / historic past—actually happened in India over the millennia not by intent or purpose, but almost entirely through natural selection, genetic and / or geographical isolation and, above all, by the survival of the fittest. This has endowed them with qualities that are unique: resistance to stress, and nutritional as well as environmental and resistance to diseases. Conservation of as many of them as is possible, if only for the posterity is, therefore, a matter of national morality.

Conservation of domestic animal genetic resources except *in situ,* among large mammals is a very complex concept, with no living examples in recent times, in any part of the world. What was economically irrelevant have vanished over a period of time, as farmers anywhere in the world would be unwilling to support domestic animal species economically irrelevant to them at their costs. Unlike in the case of plant genetic resources, preservation of a few thousand doses of frozen semen or frozen embryos, or even a few hundreds of live animals—both male and female—in a farm, cannot be construed as preservation of biodiversity.

## Rationale to conserve and develop indigenous breeds

The indigenous breeds of many developing countries fall in two major categories; those that need to be conserved to maintain biodiversity in domestic animal genetic resources and those that have the potential to produce an economic gain for the livestock keepers in a specific environment. Obviously the approaches shall also differ for the two groups.

## Breed conservation

Justifications for breed conservation can be many. Animal genetic diversity is the preserve of natural heritage; loss of a breed is an irreplaceable reduction in the natural profusion of the forms. Preservation of breeds with unique DNA will contribute to long-term research in molecular engineering. It would be a tragic commentary on mankind if, in future, when progress in molecular research opens up opportunities for selection of animals, some unique livestock genetic resources resulting from thousands of years of natural and human selection were lost. Genetic variability is the key for enhancing progress and loss of genetic variability would limit man's capacity to respond to changes in economic forces for the exploitation of animal production in tomorrow's world. Many of the tropical breeds are blessed with the ability to survive and reproduce, albeit at a low level, under disease stress and in hostile or highly specific environments. Loss of such breeds means loss of specific adaptation traits and the DNA sequence which codes for this ability. These breeds are also expected to have excellent ability to convert limited feed supply into protein. Voices opposing programmes for conservation and development are of the view that if a breed is economically useful market forces will preserve it, thereby saving the prohibitive cost involved.

Cattle breeds indigenous to the country, which provide good performance combined with adaptation to the harsh environment, including for example disease resistance, climatic tolerance, ability to use poor quality feed and to survive with reduced supplies of feed and fodder, are being neglected in the developing world. Major reasons for this negligence are: lack of awareness and inability of the small farmers who keep such animals, non-existence of breeding organisations and inertia of the breeding service providers.

## Breed development

There is an increased need for natural resistance to diseases or parasites should a current antibiotic or other treatment become unavailable or ineffective. Though there are increasing demands for milk and milk products, disproportionately higher increase in management costs raises production cost. Changes in the productivity of the breed are essential to making the enterprise cost effective and competitive. Development of local breeds for improving the economic traits without losing much of the adaptive features is possible through the time-tested methods of selective breeding and shall be taken up on a large-scale with the participation and involvement of the farmers keeping them. The genetic aspects of breed development are selection and breeding while the environmental aspects are management improvement and market. The breeding programme for breed conservation and development is similar to the one explained below and employs the general principles of animal genetics and breeding.

# Steps in implementation of a breed conservation and development programme

Breed conservation and development has a major significance in developing countries since systematic attempts have not been undertaken in the past for developing economically useful breeds. The various steps involved in a breed conservation and development programme in addition to those explained in Part II of this book are breed characterisation, breed survey and documentation, formation of Breeder Associations and Herd Book organisation.

## Breed characterisation

Only the favoured breed by farmers of a given area because of its economic values shall be considered for a development programme. The characters of the breed are to be defined, giving importance to the economic traits rather than to political and phenotypic considerations. While doing so it would be beneficial not to increase the number of breeds and not to be very strict on the phenotypic characters, as this would jeopardise the genetic progress of economic traits. Many developing countries have established National Bureaus for Animal Genetic Resources with the support of FAO.

## Breed survey and documentation

These are the first steps of conservation and development of an identified breed, which is not adequately catalogued and developed. It would be advantageous to prepare necessary formats for recording the characters of the breed before conducting the survey. The survey for all practical purposes shall be limited to areas where a sizeable number of animals of the desired breed exist. Information collected during the survey shall include:

- Number of animals according to sex and age
- Farmer rationale for rearing the animals
- Present level of production
- Means of reproduction
- Procedure used for selection of breeding bulls
- Willingness of farmers to co-operate in the breed development programme
- Special attributes to be incorporated in the selection programme

Persons trained for the purpose shall be employed for conducting the survey. A structured survey is expected to provide information about the population size, population structure, density of population, economic value of the breed and farmer preferences for the breed. Results of the survey when consolidated and analysed will give an authentic record for the breed development programme.

## Breeder Asociations

For lasting impact of the breed development programme, it has to be conceived and developed with participation of the farmers. Formation of Breeder Associations is one way through which an active participation of farmers can be achieved. Breeder Associations formed for zebu breeds with small farmers as members have to be empowered by awareness creation, activation and constant monitoring until the Association is able to run on its own. Support from NGOs and/or government will be necessary to bring the Breeder Associations up to a sustainable level.

## Herd Book Organisation

Maintaining track of individual animals with respect to pedigree, reproduction and production and accepting those which satisfy the specifications prescribed is the method adopted for maintaining purity of the breed and augmenting further progress. Individual records maintained for the purpose are called Herd Book. This system originated in the United Kingdom as early as the18th century. One of the prime jobs of the Breeder Association is upkeep of the Herd Book. Identification and registration of all animals belonging to the breed is the first step in starting a Herd Book. Details of the animal, ownership, production and reproduction data are recorded in the book in sequential occurrence. However, not all the animals would satisfy the specifications of the breed. This situation is tackled by opening two or sometimes even three registers, a primary register (sometimes even a secondary register) and a Herd Book.

## Primary Register

The Primary Register will have all animals having the phenotypic characters of the breed owned by farmers interested in continuing in the breed conservation and development programme.

## Herd Book

The Herd Book, as mentioned, is the register of animals with pedigree and breed conformation. In actual practice, Herd Book animals are first registered in the primary register and when found to conform to breed specifications transferred to Herd Book. The progeny of the Herd Book animals if born to 'approved bulls', will be registered in the Primary Register and the cycle repeated.

The registers shall be maintained with the help of a computer, which will enable quick, accurate and easy management of data. Productive and reproductive data of all the registered female stock shall be collected on a regular basis, compiled and made available to the farmers.

Individual Breeder Associations normally operate in a defined and limited geographical area. The Herd Book Organization is often designed as an umbrella organization for all Breeders Association of a particular breed. Often called Breeders Federation, this organization ensures herd book keeping and also takes cone of numerous tasks such as political lobbying, technical networking, sourcing of genetic material and conducting applied research.

## Technical approaches for breed conservation and development

The technical approaches employed to conserve and develop local breeds within smallholder production systems are the same approaches as used in commercially oriented breeding programmes, be it pure breeding or crossbreeding. To develop local breeds, normally pure breeding strategies are employed. To be successful one must be in a position to identify the best animals in terms of the breeding goals defined for the breed in question (selection). This can only be done if access to reliable data be it production, reproduction or phenotypic conformation data is ensured. Furthermore, means to propagate desired genes in subsequent generations are essential. Here again the methods employed are the same as in any other breeding programme, e.g. field performance recording, artificial insemination, progeny testing and sire evaluation techniques.

For these reasons, breed conservation and development are not referred to explicitly in subsequent chapters of this book but the methodologies and practices are absolutely valid for these programmes.

# 2. Breeding programme for large ruminants kept by small farmers

*The procedures for developing breeding programmes for cattle and buffaloes in smallholder farming systems are defined and discussed. A breeding programme has several components such as animal genetics, animal production, human resources, social and environmental science and economics. The implementation of a breeding programme for making positive and profitable changes in animal production systems can be successful only when these aspects are fully understood and suitably adapted to the given context and frame conditions. This chapter defines and develops a systematic approach to breeding programmes by clearly structuring and illustrating the various levels of the programmes. Major defined levels are the breeding policy, breeding strategy and breeding plans. This chapter sets the frame for the operational issues discussed in subsequent chapters.*

# Introduction

We have seen that for sustainability and success of a breeding programme a clear understanding of the approach is necessary. This chapter defines a procedure for developing breeding programmes for cattle and buffaloes in small farm systems. Developing breeding programmes suitable for small farmers in developing countries is not always easy because of the various elements involved.

A breeding programme has several components, e.g. animal genetics, animal production, human relations, social and environmental science, economics and behavioural science. Implementation of a breeding programme for making positive and profitable changes in animal production systems can be successful only when these aspects are fully realised and a suitable mix of the various elements is adopted.

Animal production environment can be high input, medium input or low input. The following descriptions have been given for the three types of production environments (ICAR, 1999).

High input production environment is one wherein all aspects of animal production are managed to ensure high levels of survival, reproduction and output, and production risks are constrained primarily by managerial decisions.

In a medium input production environment management of available resources has the scope to overcome the negative effects of the environment on animal production although commonly one or more inputs will limit output, survival and reproduction.

Low input production environment is one in which one or more limiting inputs impose continuous or variable pressure on livestock, resulting in low survival, reproductive rate or output. Output and production risks are exposed to major influences, which are beyond human management capacity.

Even though smallholder mixed farming production systems are generally classified under low input production environment that may not always be the case. Optimal breeding structure for small farm systems is very much determined by what is possible and what is optimal in the given situation.

A breeding programme involves all operations required to improve the genetic level of a population. It must have a strong scientific background, be cost effective and provide maximum gains for the limited resources available. It need not be characterised by latest reproductive technology and genetic evaluation software. A system to gather information on a regular basis has to be built into the structure of the breeding programme (Van der Werf, 1999). Description of the current situation, definition of the development objectives and presentation of the strategy to manage the available animal genetic resources are steps suggested by Bichard (1999) to develop enabling breeding programmes.

The steps to develop a breeding programme as described by Kropf and Chacko (1992) (Figure I.3) are employed for discussing the various aspects of a breeding programme in this book.

**Figure I.3**   Elements of a breeding programme

# Breeding policy

In a systematic policy development process, the breeding policy is part of a livestock policy, and the livestock policy is part of an overall agricultural policy.

A breeding policy is a concept for improving the genetic level of animals; it depends mainly on the biological and economical factors and defines overall objectives, means and restrictions. It must be part of the overall developmental objective of the agricultural production system. For example there is no need to increase milk production in an area when there is no demand for milk. Breeding policy has the overall goal of serving the farmers and community to establish development objectives for agricultural production with focus on judicious use of inputs per unit of output, thereby attaining sustainability.

Genetic improvement shall be in tune with farmer perceptions and wishes; this will enhance acceptability of the programme and facilitate smooth implementation of the policies. A sustainable programme converts available resources into human food without diminishing their future availability or causing environmental degradation. The policies must be governed by the requirements of the situation. In many tropical countries dairy cattle/buffalo improvement programmes are dependent on the gene pools of exotic dairy breeds, zebu breeds and dairy buffalo breed/s of the region.

## Breeding policy decisions

While developing breeding policies for dairy animals in developing tropical countries, the flow diagram (Cunningham and Syrstad, 1987) given in Figure I.4 will be useful.

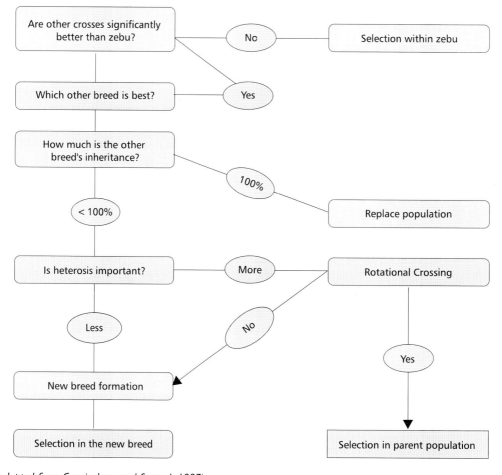

*(adapted from Cunningham and Syrstad, 1987)*

**Figure I.4**  Strategic option involving crossbreeding and selection

The objective of a breeding policy has to be robust while simultaneously considering economic and social developments. Development of breeding policies is a process that requires continuous refinement after initial planning and during implementation (Figure I.5). For a given population the breeding policy is developed by applying the science of breeding, the experience gained to date in the area, duly considering farmer needs, market demands and applying management conditions prevailing in the area.

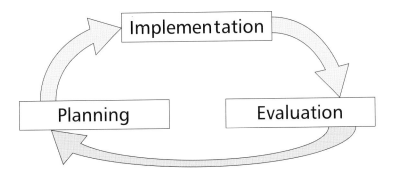

**Figure I.5**  Continuous nature of developing breeding policies

A review of the breeding systems in developing countries revealed that economic, social and ecological production environments influence them. A short description of the evolution of breeding systems in India is given in Box I.2.

## Examples

### Crossbreeding

The percentage of cattle of recognised zebu breeds among the total zebu cattle in the developing countries of Asia and Africa is low. The majority of the cattle in these regions conform to no recognised breed and are generally referred to as non-descript local. Crossbreeding with any of the known dairy breeds of the world and limiting the level of donor inheritance at a certain stage is attempted for milk production improvement in subsequent generations.

An illustration of a breeding policy wherein the level of exotic inheritance is maintained at around 50% is given in Figure I.6. Though the level of inheritance of the crossbreds in the $F_1$ generation is equal to half that of the parent breeds, it will not however be equal to half that of the parents in the subsequent generations due to increased chances of linkages and crossover of chromosomes. As such we can only say that the level of inheritance has approximately the desired proportion.

*The Sunandini breed was developed from the local cattle of Kerala, India; a Sunandini cow.*

There are two phases at the policy level:

- The crossbreeding phase in which the desired level of exotic inheritance is brought in through one or more forward crossings and
- The selection phase wherein selected crossbred bulls of the desired 'donor inheritance' are used for mating. The crossbreeding phase will extend to more than one generation when the exotic inheritance level is more or less than 50%.

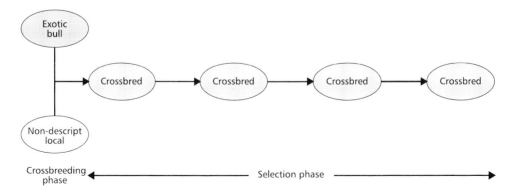

**Figure I.6**  Crossbreeding programme limiting 'donor inheritance' to around  50%

**Box 1.2**  Evolution of breeding systems in India

In India the following breeding practices were followed for a long time (Intercooperation, 2000).

**Indiscriminate breeding.** This is a traditional practice farmers continue to use. Essentially, animals are bred or allowed to breed indiscriminately following no specific breeding plan. Mating is more a matter of chance, depending on availability of bulls/bucks in pasture or in the village. The farmer's main goal is to get the females pregnant in order to get an offspring and/or a lactation. This is also linked to the common herding systems on common property resources and requirements of the cattle owner: social standing, draught, dung and milk.

**Pure breeding in specific breeding tracts.** The existence of clearly defined Indian cattle and buffalo breeds is the result of selective animal breeding, which has been practised over centuries. Livestock holders consciously maintained certain breeds by selecting for those traits that were important for their particular production and socioeconomic environment. Selection within local breeds is still going on, but not as systematically and scientifically as it could be. Field performance recording (FPR) and progeny testing (PT) in future will be the tools to obtain improved production in local breeds and thereby facilitate maintaining large biodiversity in Indian indigenous livestock. In pure breeding of local breeds there is always a dilemma between conserving biodiversity of domestic animal breeds and the undisputed need to achieve an acceptable economic return.

**Crossbreeding to enhance milk production.** Crossing of local female cattle with exotic males began on a large-scale during the sixties. This was initially done by natural service with imported animals and later through artificial insemination first with liquid semen and later with frozen semen. Under favourable conditions, crossbreeding resulted in a two- to threefold increase in milk yield of the resulting progeny. Systematic crossbreeding requires the production of crossbred bulls for inter se breeding. Systematic selection within the crossbred population (FPR, PT of bulls, nominated service) results in a population which fits best into the specific environment and the initially defined blood levels of the crossing partners is no longer important and can no longer be exactly determined. Such a programme will lead to the creation of new breeds, made to measure to a specific condition, provided selection is done on the basis of reliable field performance data and a sound breeding programme is followed over several generations.

**Upgrading to enhance milk production.** This practice was mainly carried out in non-descript buffaloes with Murrah buffaloes and in non-descript cattle with Indian draught or milk breeds. This system of mating is still being practised, possibly to a lesser extent in cattle. Upgrading takes place when crossbreeding programmes fail to produce good quality crossbred bulls for inter se mating. In such cases, genetic progress is often continuously imported and the programme evolves from crossbreeding to pure upgrading. With increasing exotic blood levels, the adaptability of the animals to the environment and their suitability to the smallholder fodder base often decline. Lately, the upgrading of descript cattle breeds with exotic breeds to reach almost 100% of exotic blood level has also been observed in areas with specialised milk production units.

**Mixed breeding systems.** There is no single ideal system for an area or region. A mix of systems will be required to suit the diverse needs of individual production systems.

### Selection within breed

Selective breeding is the policy for breed conservation and development of zebu breeds (Figure I.7). The basic principles of genetic selection must be applied here. A herd of superior animals can be maintained by AI organisation or selected from a group of cows belonging to farmers controlled under field performance recording. Breeding bulls for artificial insemination and/or natural service can be produced from such herds.

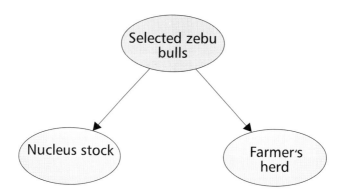

**Figure I.7**   Selective breeding for genetic improvement within a breed herd

### Upgrading

The target herd is continuously mated with bulls of the desired breed, resulting in a gradual transformation of the target herd to that of the donor herd. The system of upgrading is described in Figure I.8. Genetic improvement of buffaloes, which form the mainstay for milk production in many developing countries, is possible by continuously upgrading them to any of the desired dairy buffalo breeds accepted by the farmers.

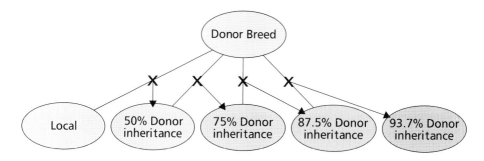

**Figure I.8**   Continuous upgrading

# Breeding schemes

The breeding scheme is the second step in discussing a breeding programme. It defines the structure of the population, the gene flow from one generation to the next and the selection methods for each path.

## Population structure

A group of animals sharing a common gene pool and sustaining themselves through interbreeding is called a breeding population. It can be seen that the population covered under a breeding programme is stratified, comprising a nucleus herd having genetically superior animals, field performance recorded herd, also known as the multiplier herd, and target population. The nucleus herd is responsible for most of the genetic changes, since they are used for the production of breeding bulls. A group of animals constantly under field performance recording (FPR) and maintained by the farmers is called the multiplier herd. The superior female stock of the multiplier herd is elevated to the nucleus herd when they are selected as dams for the next generation of breeding bulls. The target population is at the bottom of the 'population pyramid' and is the benefactor of the changes brought about through breeding. The composition of a breeding population is described in different levels constituting a pyramidal structure with genes flowing among them. (Figure I.9).

Separate breeding schemes are necessary for different subpopulations. In a population consisting of non-descript cattle, a defined zebu dairy breed and local buffaloes there will be three breeding policies and three breeding schemes.

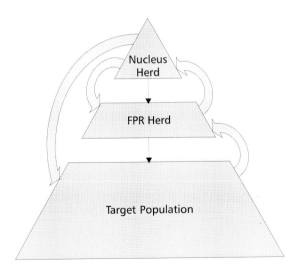

**Figure I.9**   Structure of breeding population

## Gene flow

The gene flow calculates the genetic response and specifies its dynamics by describing the paths of flow of genes through populations and through generations. Genes flow from one generation to the next through the paths sire to son and daughter as well as from dam to son and daughter. In actual practice however, there are various subgroups of animals in each of the paths depending on the type of selection applied. Box I.3 gives an example of the various groups of animals to be considered when non-progeny tested bulls are used for artificial insemination and proven bulls for production of the next generation bulls and in which the population has a nucleus herd and a target herd.

**Box I.3**   Various groups of animals to be considered in breeding programmes

---

**Proven bulls to produce the next generation bulls.** These are proven bulls and their genetic superiority depends on the accuracy of progeny test and the intensity of selection. They are mated to elite cows to produce the next generation bulls.

**Breeding bulls used in large-scale AI.** The genetic potential of the bulls used for artificial insemination will be related to the breeding value of their sires and dams.

**Young bulls under progeny test.** These are bulls recruited to the programme and put to the progeny test regularly. The number of young bulls tested depends on the type of selection applied. When all young bulls tested are selected for the artificial insemination programme the number is equal to the replacement requirement and when 50% are selected for artificial insemination the number of bulls tested will be twice the number of bulls required for replacement.

**Bull mothers.** Cows identified as mothers of future bulls in the nucleus herd either in the farm or in an FPR area or both and inseminated with the semen of bulls selected as sires of the next generation bulls.

**Heifers in the nucleus herd.** The adult females born out of the mating between bull sires and bull dams in the normal course form the replacement for the bull mothers.

**Target females.** Adult females in the area covered by artificial insemination are referred to as target females and are inseminated with the semen of bulls selected for artificial insemination.

---

Information on the prevailing animal production systems is necessary for deciding the breeding scheme. A scheme wherein the genetic improvement flows from one generation to the next through selected males is illustrated in Figure I.10. The breeding males are produced either in the bull mother farms of the artificial insemination organisation or in the field performance area or a combination of the two. The breeding scheme depends on the breeding population and method of selection.

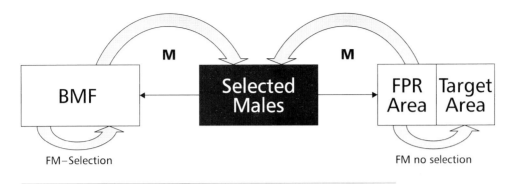

FM-Selection                                    FM no selection

BMF: bull mother farm, FM: female, M: male, FPR: field performance recording

**Figure I.10**   Breeding scheme—Pure breeding

## Selection methods

The information on which selection is based is referred to as method of selection. In other words, it is the method of estimating the breeding value. There are four common methods of estimating breeding values.

| | |
|---|---|
| **Individual selection:** | Individuals are selected based on their own phenotypic values |
| **Sib selection:** | Individuals are selected based on the performances of their sibs |
| **Progeny test:** | Individuals are selected based on their daughter's performance |
| **Animal model:** | Individuals are selected based on their own production records and the breeding value of their relatives. |

The selection methods considered in a breeding scheme have to take into account traits to be selected, infrastructure facilities that can be made available and types of activities envisaged.

Selection of traits in a breeding scheme is related to the needs of the customer. A large proportion of small farmers in developing countries rear animals as part of the small-scale crop livestock mixed farming system. A proper understanding of farmer needs is necessary before elaborating the breeding scheme. Gathering reliable information from the farmers is rather difficult in countries where sociocultural background, literacy rates and awareness act against correct information gathering. In such situations a properly designed survey conducted by specially trained personnel together with case studies will provide the relevant information.

As a general policy, breeding schemes are designed for increasing milk production and very little importance is given to meat production or draught power in many of the small farmer production systems. The need to increase milk production should be challenged with the cultural background of the area, farmer habits, how fast these habits may change, importance of other occupations and the returns expected from dairying. This need not be the case for all the situations. Sustainable cattle development programmes cannot be implemented for the farmers; they have to be conceived, planned and executed with them (Menzi, 1993).

Infrastructure facilities made available for implementation of a breeding programme vary from place to place. Analysis of many breeding programmes in developing countries reveals that they were designed for optimum conditions and could not be successful in less optimum situations (Tewolde, 1999; Taneja, 1999). Success of a breeding programme will depend on the use of a condition friendly programme. It is always possible to upgrade the programme in line with improvements taking place. In summary, what is required is a programme which is in harmony with the facilities available, with the wishes and likes of the farmer, with the prevailing market situation, with the existing sociocultural and agroeconomic background and functional style of the breeding organisation in place.

# Breeding plan

## Developing a breeding plan

The third step is the elaboration of the breeding plan according to the defined breeding scheme. It is developed for a certain time period. For large ruminants like cattle and buffalo, breeding plans are drawn up for a period not less than 5 years. It indicates the number or percentage of animals born, identified, lost, bred, evaluated and selected in a breeding scheme. It is further elaborated on the means of breeding (artificial or natural), the requirement of bulls, semen, their placements and movement. The breeding plans have to be elaborated for each subpopulation separately. Steps in the development of a breeding plan are described in Figure I.11.

## Quantification

Following the flow chart given in Figure 1.11 a series of steps are described (Chacko, 1999) in Figure I.12 to quantify the breeding plan. Using simple logic

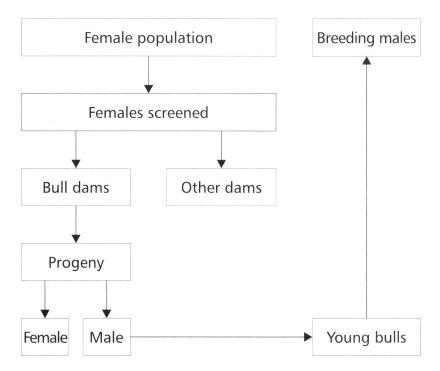

**Figure I.11**   Steps to develop a breeding plan

estimates are made of the numbers of annual artificial inseminations, frozen semen doses to be available, breeding bulls and bull dams.

Simple logic is also employed to develop the steps in Figure I.12 to illustrate the principles. While the rate of mortality and culling among cows would pull down the estimated figures, fresh females coming into production would push it up. It is not attempted here to further complicate the calculations by adding these aspects since for all practical purposes the steps shown in Figure I.12 would provide a reasonably good estimate. A formula developed from the steps given in Figure I.12 to calculate the number of artificial inseminations to be carried out in a population is given below

Number of AI per year (AIY) = BFP · COV · AIC ·12 /CI
Where: BFP is the total number of adult female population;
COV is the proportion of animals covered by the programme;
AIC is the average number of AI required to produce a calf;
CI is the average calving interval in months.

This formula is further extended to calculate the number of bull mothers to be maintained in the bull mother farm of the artificial insemination organisation and is given in Box I.4.

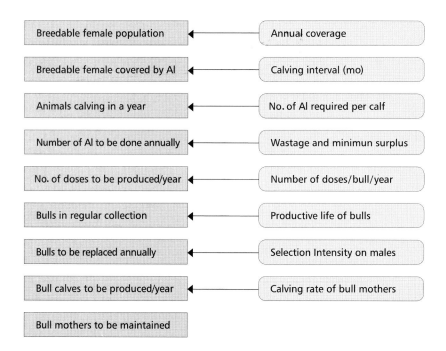

**Figure I.12** Steps to quantify a breeding plan

**Box I.4** Steps to calculate the different figures in a breeding plan

| Details | Symbol | Formula |
|---|---|---|
| Estimated number AI per year | AIY | |
| Expected wastage of semen doses & minimum surplus planned in the AI organisation (in proportion) | SUR | |
| Doses to be procured/produced annually | p | AIY(1+SUR) |
| Estimated percentage of semen doses produced by the AI organisation out of the total requirement | SPAIO | |
| No. of doses /bull/year | DBY | |
| Bulls in regular collection | p/DBY | |
| Productive life of bulls | BPL | |
| Bulls to be replaced/y | p/DBY*BPL | |
| Bulls procured from outside | BOS | |
| Bulls to be produced by AI organisation | q | (p/DBY*BPL) - BOS |
| Selection intensity on male (%) | SI | |
| Bull calves to be produced/y | q*100/SI | |
| Calving rate of bull mothers | CR | |
| Bull mothers to be maintained | | q*10000*2/(SI*CR) |

Exercise 1 of the toolbox given in the attached CD gives an option for the reader to estimate the annual artificial insemination figures for specific situations. The reader may type in the variables appropriate to her/his situation.

*Exercise 1.    Estimation of total AI in a given situation.*

The breeding plan has to take stock of the available infrastructure with respect to artificial insemination units, semen production stations, semen distribution centres and liquid nitrogen production/procurement facilities. This is necessary to plan for additional requirements.

## Selection plan

Selection brings in genetic progress and is the result of accumulation of additive effects from generation to generation. In practice it generally delivers genetic gains in line with the theoretical expectations (Smith, 1984). Plans are to be developed separately for the selection of bull mothers and bulls. It is a common practice in many developing countries to maintain bull dams by the artificial insemination organisation. It is possible to have bull mothers registered with the farmers, however, in areas where a field performance recording programme is functional. Selection of bulls for characters not expressed by them is based on the performance of their maternal ancestors, half-sibs or daughters. Further details of the selection programmes are discussed in Part I, Chapter 3.

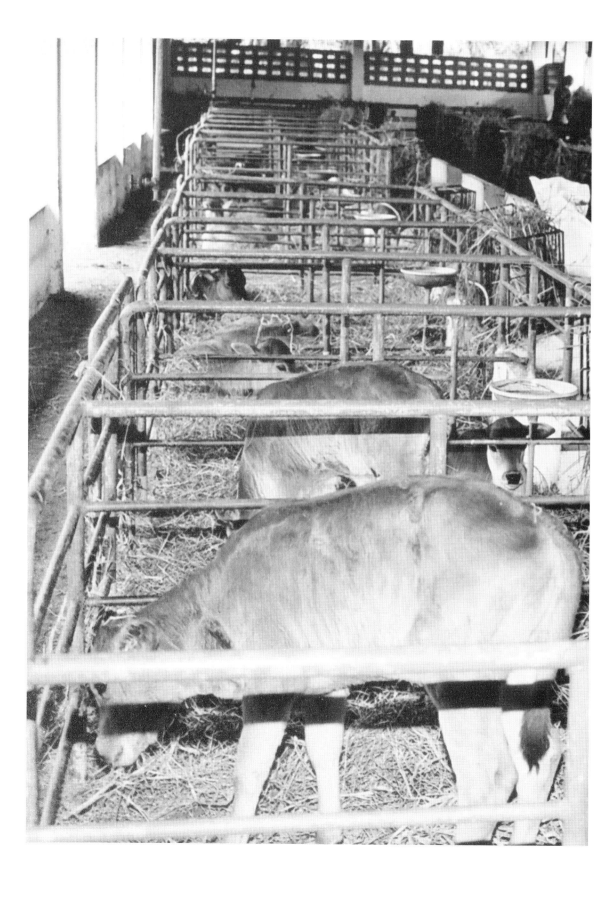

# 3. Production of breeding bulls

*The various operational aspects of breeding programme implementation are defined and discussed in a series of chapters starting with the production of breeding bulls. The number of offspring born for males are highter than that for females when artificial insemination is used as a means for genetic improvement. As such, large ruminant breeding programmes always focus on the selection path bulls to breed bulls and cow to breed bulls thereby realising a comparatively high genetic gain from one generation to the next. The procurement of bulls from various sources is discussed focusing on the rearing of bulls within a given organisation. The purpose and the management of bull mother herds are described in detail, including calculations on the size of the herds and estimation of important management parameters.*

# Introduction

There are various activities under breeding operations and are discussed in the next five chapters of this book. Management of bulls is logically the first in the series. A large proportion of the semen doses needed for artificial insemination as a general rule is produced by the organisation responsible for the implementation of the breeding programme in a given area. They maintain breeding bulls for the purpose. Production and procurement of the required number of bulls as per the breeding plan are discussed in this chapter. Breeding bulls are obtained through any one or a combination of the following methods.

- Import
- Purchase from outside sources
- Production using farmers' herds in the breeding area and
- In-house production

# Import of bulls, semen and embryos

Exotic bulls are required for crossbreeding and/or for bringing in superior germ plasm for breed development. The breeds to be imported are decided once the breeding policy is finalised. If some of the exotic breeds recommended in the breeding policy are already available within the country, a decision on import is taken based on the numbers available and their quality. Exotic germ plasm is imported as live animals, frozen embryos or frozen semen. Import of live animals enables faster use of the germ plasm in the breeding programme, while importing embryos and or semen is easier and cheaper. There are regulations for the import of animals/ embryos/ semen and they vary from country to country and are generally based on health and sanitary regulations. The ministry/department responsible for issuing licenses to import livestock will provide the required information.

There are large differences in breeding value estimations and its expression between countries. A careful examination of the breeding value estimates supplied by different countries is necessary to obtain comparable figures. While effecting import it is advantageous to get offers from trading companies for the supply at the disembarkation airport/seaport inclusive of all costs like taxes, levies for health examinations, import taxes, transport charges, loading and unloading and clearing charges at the port of disembarkation.

# Bull purchase

Pedigreed herds belonging to farmers within the country but outside the breeding area are another source for breeding bulls. Bulls of indigenous breeds of cattle and buffaloes are to a large extent purchased from such sources. It is preferred to purchase breeding bulls for artificial insemination at the age of one year and in any case before they are ready for semen collection training. This will enable the semen production station to train the animal to their requirements.

The milk production potential of the animal can only be assessed based on the milk yield of their mothers in the absence of reliable information about its breeding value. When regular milk recording is not conducted, which is often the case with many private farms, recording the milk yield of all the animals for three consecutive milkings, (evening, morning and evening) and taking the sum of the second and third records will give an estimate of the daily milk yield of individual animals in a herd. The average of this is the daily shed average or herd average. Individual cow's milk yield as a deviation from the herd average corrected for stage of lactation can be used as a measure of the genetic potential of the animal.

# Transport of animal

Special care must be given while transporting animals especially young calves. The selected animals must be treated for external and internal parasites one week before the date of transport. Safe loading, quality of the truck, care during transport, timely watering and feeding and careful offloading must be ensured. Transport should be arranged when the atmospheric temperature is optimum. A detailed checklist for transporting animals is given in Box I.5.

**Box I.5**   Aspects to be considered while transporting young males

- The trucks must have the required license and permit to transport animals
- The vehicle must be safe and allow for adequate space and ventilation
- A bedding 5 cm thick made of soft and water absorbent material has to be provided
- Limit the speed for the comfort of the animals
- Avoid driving during the hot time of the day
- Care is required while loading and offloading the animals

Contd.

---

**Box I.5** Contd.
- A first-aid kit must be carried to provide any emergency health care
- During long trips the animals must be periodically checked for injury or illness
- Veterinary care for bulls in transit has to be provided as and when necessary
- Watering and feeding must be done once in every four hours
- A responsible person should accompany the animals in the truck
- Persons transporting bulls need to be knowledgeable about the basic care of bulls
- For long-distance transport the animals should be insured for transit accidents and death
- At the destination the animals should be offloaded at a comfortable place
- Imported animals must have health certificates and be kept in quarantine as per regulations of the country of entry

---

# Bull production in the breeding area

## From farmer herds

Data available from the FPR will help to identify high-yielding cows owned by farmers. Experience in India shows that such herds are superior to those maintained by the artificial insemination organisation (AIO). These cows designated as the bull mothers in the field are mated to top bulls available with the AIO. The male calves born from such matings are purchased by the AIO and brought up as replacement young bulls.

A programme for production and procurement of young males from farmer herds can be designed through an FPR programme for which the following steps are necessary:

- Locate suitable area/areas for field performance recording
- Register and identify the animals
- Establish an FPR programme for the traits decided
- Evaluate the herd for the traits to be selected
- Identify the elite cows/she buffaloes as bull mothers
- Arrange for nominated mating of these animals with selected bulls
- Follow up the male calves born
- Decide on the age at which the calf is to be purchased
- Examine the young male calf for phenotypic characters, legs and health
- Fix the price for the selected male calf
- Arrange transport to the rearing unit

Details about organising an FPR programme in small farmer herds are discussed in Part I, Chapter 4.

*Efforts to conserve and develop local breeds were taken up in India in recent years: An Ongole bull used in AI*

## In-house production

Replacement requirement of breeding bulls can be met partly or completely from the organisation's own farms. The following aspects must be finalised while developing the breeding plans:

- Number of bulls to be replaced annually
- Proportion of the annual replacement of bulls to be produced in the organisation's farms
- Selection intensity to be applied on the dams of these bulls
- Herd strength of the farm
- Selection intensity and reproductive efficiency of the female stock

Exercise 2 in the toolbox of the attached CD will facilitate the reader in calculating the herd strength of the bull mothers farm of the AIO under a set of assumptions.

---

*Exercise 2.    Herd strength of the bull mother farms of the AI organisation*

---

Semen from bulls having high breeding value has to be employed for the nominated mating in the bull mother farms. A mating programme that helps to minimise inbreeding and to maintain a wide genetic base in the ensuing generation has to be implemented. The various aspects of management of bull mother farm are discussed below.

# Bull mother farms

The artificial insemination organisation establishes new farms or existing farms are restructured for the production of bull calves in countries where large privately owned farms are not many. Optimum management of a bull mother farm has to be ensured for getting good quality bull calves. The manager of the farm in consultation with other employees has to prepare a calendar of various daily and periodical activities and engage the required manpower to carry it out. The various topics discussed in this chapter are herd strength, selection procedures, calf management, management of growing stock, management of cows and estate management.

## Herd strength

The estimated number of animals in the bull mother farm based on the annual replacement requirement of young bulls can be worked out using Exercise 2 of the toolbox. Maintaining herd strength with the elimination of animals that are substandard in production is possible when the reproductive efficiency is optimum. It can be shown that under optimum reproductive management 33% of the cows can be culled for substandard milk production after completion of the first lactation without affecting herd size. Selection intensity that can be obtained in a herd of bull mothers can be calculated using the following formula (Chacko, 2003).

$$GS = \frac{A - (0.5A \cdot RES \cdot SUR)}{0.5A[(SUR \cdot RES)^2 \cdot CIN + (SUR \cdot RES)^3 CIN^2 + \ldots + (SUR \cdot RES)^n \cdot CIN^{(n-1)}]}$$

Where: GS is the genetic selection after first lactation in proportion to
A, the number of heifers in the farm;
SUR, the survival rate—the proportion of animals alive of the total number
of female calves born;
RES, the reproductive success—the proportion of animals conceived
during each reproductive cycle. The remaining animals are culled and
removed from the herd.
CIN is the calving ratio—the ratio between 12 and the average intercalving
period (months) of the farm.
N, the average number of lactations for which the cows are kept on the farm.

The selection intensities that can be achieved with varying levels of reproductive
management calculated using the above formula are given in Table I.4

**Table I.4** Selection intensity obtained at varying levels of reproductive management

| Average calving interval (mo) | Reproductive success (RES) | Number of lactation kept (n) | | | |
|---|---|---|---|---|---|
| | | 4 | 5 | 6 | 7 |
| 13 | 0.85 | 0.86 | 0.73 | 0.66 | 0.61 |
| | 0.90 | 0.70 | 0.58 | 0.52 | 0.47 |
| 14 | 0.85 | 0.98 | 0.85 | 0.78 | 0.74 |
| | 0.90 | 0.81 | 0.69 | 0.62 | 0.58 |
| 15 | 0.85 | — | 0.98 | 0.91 | 0.87 |
| | 0.90 | 0.91 | 0.79 | 0.73 | 0.69 |

(Source: Chacko, 2003)

In a herd in which the cows on average are kept for 6 lactations and with 85%
success in reproduction, the percentage of cows that can be selected for milk
yield is 66 and 78 respectively with average calving intervals of 13 and 14
months. It can be seen from the table that with longer calving intervals and
fewer lactations, selection intensity decreases. With an average calving interval
of 15 months and cows kept for 4 lactations and RES of 0.85 no selection is
possible and the herd size will decrease year after year. In this formula a constant
mortality rate of 5% is assumed. By storing the formula in a computer and
keeping the option to change the variable according to the actual situation of
the farms, it is possible for the manager of the farm to forecast the extent of
selection intensity possible in a given situation. A model to calculate the
selection intensity with changing parameters is given in the toolbox of the
attached CD as Exercise 3. It also gives an indication of the total number of
female stock in the farm at a given time.

*Exercise 3.    Selection intensity for milk yield after first calving in a bull mother farm*

**Calculation of herd strength**

It is possible to calculate the herd strength of a farm consisting of cows, heifers and female calves using the same model. The number of cows to be kept in a farm to produce a fixed number of heifers depends on the survival rate among female calves, reproductive success and the calving interval. Unwanted animals (poor reproducers, poor producers and debilitated and weak animals) must be culled and removed from the herd at regular intervals (at least three times a year). Powers to take culling and disposal decisions have to be vested with the manager of the farm.

## Selection procedures

Good cow-culling decisions require ongoing month-by-month evaluation and cow comparisons. It is recommended that these decisions be based on both the animal's present and anticipated future profit. The disposal of culled stock must be planned to minimise extended low-profit periods. Genetically superior cows must be fed and managed to obtain maximum production. Selection of cows is to be done based on milk production performance. For making unbiased comparisons between cows in the herd, lactation yields have to be corrected for environmental influences such as effects of year, season of calving, age at first calving and parity. A simple method to calculate the breeding value of cows is explained in Part II, A, chapter 4.2.

For developing accurate correction factors sophisticated statistical methods such as least squares analysis and best linear unbiased prediction are applied. A procedure has to be developed for such type of analyses periodically. By keeping the correction factors updated as and when they are modified in a subfile of the software package, the computer can calculate the breeding values periodically.

## Calf management

Optimum management is essential to ensure proper growth of the calves. Calf identification, feeding, health care and growth monitoring are the major aspects in calf management.

## Identification

Calves must be identified with a permanent number soon after birth and a herd book opened for the calf with details of pedigree, birth date, breed, sex, birth weight and other relevant information if required. Ear tags, plastic or metal, are found to be good. Electronic identification systems are currently available in the market for identification of animals and record-keeping. When supported by computer packages record-keeping and management-monitoring become easier.

## Feeding

Feeding of young calves and growing stock needs special attention. Newborn calves must be fed colostrum soon after birth and thereafter milk at the rate of 10% of the live body weight of the calf. Milk feeding can be substituted with milk replacers. The feeding of milk/milk replacers has to be continued until the calves are three to four months of age. Good quality hay and tender greens must be made available in the manger for the calf to nibble *ad libitum*. Young calves start consuming concentrate cattle feed at the age of 30 days. The concentrate feed given to calves must have a digestible protein content of around 20%. There must be a regular supply of good quality water accessible to the calves at any time of the day. Automatic water dispensers attached to the individual pens are found to be very effective. The faster the calf grows, the earlier it exhibits sexual maturity and the earlier the bull can be proved.

## Housing

Young calves are best brought up in individual pens until one year old. Maintaining calves in individual pens facilitates:

- Controlled and individual feeding
- Monitoring growth
- Checking calf scour and other infections
- Prevention of testicle sucking by calves

Keeping calves in individual calf pens proved advantageous for faster growth, less testicle damage due to sucking by other calves and monitoring growth. Calves kept in individual pens or tied have to be sent out daily to an open paddock for about an hour for exercise and grazing.

## Health care

Infection with external and internal parasites is often an important threat to the health of the calves. Anthelminthic treatment and protection from external parasites must be a regular practice. Two of the common diseases affecting young calves are calf scour and pneumonia. Maintaining hygiene in the sheds and timely treatment are necessary to prevent calfhood diseases. Calves must be vaccinated against the contagious diseases prevalent in the area at the appropriate time. Many managers practise dehorning of the calves. Caustic chemicals or electricity is used to remove the horn buds before the calves are one month of age. While dehorning is said to give the herd a uniform appearance, it can destroy the individuality of the animals, breed characteristics and the real life appearance of the herd. There is evidence of polled bulls being extremely dangerous in open yards in Australia (Lamond and Campbell, 1970).

### Growth monitoring

Repeated recording of weight and measurements at regular intervals are useful in monitoring the growth of young calves. Calves born in the same week can be weighed and measured on the middle day of the week. Weighing shall be carried out once in a month till they are 12 months of age. Keeping a live body weight graph in front of each calf will help the manager to detect poor growing calves and take corrective measures.

### Selection

Male calves to be raised as future bulls are initially selected based on the performance of their parents, physical fitness and growth. The cut off limits with regard to milk yield or breeding value of the dams should be decided in advance. These selection limits are to be revised on an annual basis. Selection for growth, physical fitness and genetic qualities must start at the calfhood stage based on set standards. It is also equally important to dispose of the unselected soon after taking the selection decision so that better care can be given to the selected stock. Young calves must also be culled for physical abnormalities and chronic diseases.

## Management of growing stock

### Housing

Calves above one year of age are generally kept together in a paddock until they attain maturity. The design of the shed must be suitable for the climate and management situation. There must be sufficient ventilation and protection from heat and cold. Sheds with partly open and partly roofed area are recommended for hot and humid climate. The flooring must be concrete, stone/brick laid or soil depending on the management practice adopted. Male and female calves have to be housed separately after about nine to ten months of age.

### Nutrition

Growing stock must be maintained on a high roughage ration with supplementation of minerals and salt. In places where grazing is possible a large proportion of the nutrient requirement must come from grazing. Clean water must be available in the paddocks. Animals overfed in early life also have poorer reproductive performance in both their first and later lactations and produce less lifetime milk.

### Health care

The animals must be vaccinated against the contagious diseases prevalent in the area and protected from parasitic infestations by treatment with anthelminthics, sprays and dips. Unhealthy animals in the group must be separated and given special care.

### Reproductive management

Sexually mature male calves are transferred to the semen production station for training and induction into the group of regular breeding bulls. Young females ready for artificial insemination are to be separated, watched for good heat symptoms and inseminated at the right time. It is not uncommon that up to 10% of heifers fail to breed. Poor growth or obesity will increase infertility and the dropout rate among heifers.

## Management of cows

### Nutrition

Feeding is the single largest item influencing the production and productivity of the animals and in India accounts for around 55% of the total cost of milk production. Scientific feeding is important for optimum production and is essential for making correct selection decisions. Depending on the availability and quality of arable land, the quantity of roughage that can be produced on the farm has to be estimated at the beginning of the year. The estimated deficit of nutrients has to be covered by concentrate feeds. Aspects such as nutrient requirement, computing balanced rations, mineral and salt requirements, and various cattle feeds for cattle are not discussed in this book. The reader may refer to any standard textbook on nutrition for this. The cost of concentrate feeds can be considerably reduced if they are purchased and stored during the flush season.

*Feeding concentratres is very popular in Kerala. An elite cow belonging to a farmer.*

## Housing

Clean and comfortable housing helps to increase production and life of the animals. Sheds with two rows with face-to-face arrangement are considered efficient. Front passage, manger, standing space, urine and dung channel, and rear passage are the different sections of a cowshed. A front passage of 3 m width will allow a tractor to go inside the shed and offload the feed and fodder. The manger has to be 55 to 65 cm wide with its outer edge level with the front passage to facilitate easy transfer of feed and easy of cleaning. The standing space has be about 10 cm lower than the manger level. For each animal 85 to 100 cm width must be provided. The length of the standing space is to be adjusted according to the average size of the animals to be stalled and varies from 1.8 to 2 m. A gradient of 1 cm in 100 cm from the front to the back of the standing area is to be provided for drainage of the urine and shed washings to the dung channel. The dung and urine channel must be 10 cm deep and 45 to 50 cm wide and sloping to the slurry tank end of the shed. The milkers and cleaners often use the rear passage for which a width of 1.2 to 1.5 m is necessary. Roofing can be tile, concrete or asbestos sheets. A half-wall 1.0 m high on the rear side helps to prevent strong wind entering the shed. Other important constructions in a bull mother farm are: slurry tank, hay store, silage pits/silos, feed store, implement shed, health care room, milk room, garage, manager's room and residences for employees.

## Milking

Milking is one of the most important activities in a dairy farm and needs trained hands. Milking at five in the morning and five in the evening keeps the interval between two milkings equal and makes daytime milking possible. One milker can hand-milk 12–15 cows in addition to removing the dung and feeding concentrates. The milkers must be trained in the correct procedure as well as the hygienic practices in milking. If milking machines are employed aspects such as cleaning the machine, adjusting pulsation of the machine, regulating the vacuum and availability of electricity are important. When milking machines are used, the number of cows that can be controlled by one person increases considerably.

## Health care

Health problems increase the rate of involuntary or forced culling, which results in reduced selection intensity. Each high-yielding cow culled involuntarily due to health problems imposes retention of a cow inferior in milk production. Infertility, mastitis, udder problems, other diseases and injuries affect good and poor cows in the herd with about equal frequency. Large herd-to-herd differences in the rate of involuntary culling emphasise the importance of good herd management to keep the number of involuntary culls minimal. Prevention through management and routine veterinary service, rather than using veterinary service only for emergencies must be the rule. Individual health records are an essential part of these programmes and are helpful to the herd manager, veterinarian and AI technician in implementing the most effective practices for profit maximisation.

## Record-keeping

Good records will help the dairyman exert tighter control over management inputs. Information on growth, milk production, health, occurrence of heat, calving and feeding, are important for deciding when and which cows to watch for heat, pregnancy check, to cull, to dry, to prepare for calving, as well as how much to feed. Production records are as fundamental to effective management as they are to genetic appraisal. The management has to develop a set of formats and registers to record the required information. It is not uncommon that many bull mother farm maintain more registers than required giving rather less importance in its application in management and selection decisions. There are also very user-friendly and efficient software packages for the management of the bull mother farms. Box I.6 gives a list of registers generally maintained in a bull mother farm in India.

**Box I.6**  Common registers maintained in bull mother farms in India

**Livestock Register** to record all details of the animal and to serve as the stock register. Recording the information systematically and regularly must be ensured.

**Calf Register** to record the details of the calves born.

**Daily Milk Recording Register** to record the daily yield of individual cows must be maintained up to date. At the bottom line of the register the total milk produced in a day must be recorded by adding the individual cow's yields.

**Milk Disposal Register** helps to regulate the disposal of milk from the farm on daily basis. It is desirable to have this register kept by an officer other than the milk-recording officer. It must have provisions to record the receipt of the milk, quantity given for feeding the calf, sold to the employees of the farm, sold to milk society, sold to others and the closing balance.

**Feedstock Register** to account for the concentrate cattle feed used in the farm. It must have separate columns for recording the stock of various types of feed items and show the daily release of feed from the store. The number of columns will be regulated according to the number of feed items to be procured in the farm, giving 3 columns (receipt, issue and balance) for each item.

**Artificial Insemination Register** has records of AI done in the farm. This is a primary register and the artificial insemination particulars pertaining to individual cows. The artificial insemination register must have columns for the date, bull number, result of AI and a remarks column to write any extra details as and when necessary.

**Weight Recording Register.** The live body weight and the body measurements (heart girth circumference, height and length) are recorded in this register.

**Pedigree Card.** For a quick look at all the information of an animal a single sheet pedigree card is maintained.

**Health Register** is a routine casebook in which all treatments and vaccinations given to an animal are recorded. A separate format may not be necessary for a health register.

### Reproductive management

Efficient reproductive management is one of the most important elements in management of cows. It has a major effect on the profitability of the farm and will help in increasing the selection intensity there by improving productivity over generations. Unlike genetic progress, improvements made in reproduction during one year are not carried forth to the next year unless improved practices are maintained. Because a missed heat is more costly than a low conception rate, attempts to breed the animals at the earliest opportunity are very important. When all economic factors are considered, improved heat detection has a large beneficial economic impact. Improving heat detection to 80% is worth a significant investment in detection procedures. Missing an opportunity to inseminate means an increase in the calving interval by 21 days. The duration of oestrus is shorter and the manifestation of heat symptoms is less pronounced in hot and humid climates than in temperate climates. Further the zebu breeds are generally shy breeders.

Maintenance of high fertility will lead to production of maximum milk in a lifetime as well as maximum number of offspring. Maximum selection can be exercised when the first calving age and intercalving period are reduced to the minimum. The farm managers should set standards for the various reproductive parameters. Some of the measures to be taken for optimum reproduction of cows and buffaloes are:

- Care and feed female calves properly to achieve puberty and the required live body weight at the earliest age
- Avoid post-partum infection in cows
- Make the herd's men aware of the importance of correct reproductive management
- Practice correct technique of artificial insemination using good quality semen
- Check non-repeaters for pregnancy preferably within 45 days of artificial insemination
- Examine and treat the repeat breeders
- Eliminate chronic non-cycling and problem animals rather than treating for long
- Adopt correct feeding of animals to maintain an optimum reproductive status
- Supplement trace elements as required
- Keep regular and correct records of reproduction

**Percentage of cows in milk**

In a herd of cows with an average calving interval of 12 months, 100% of the cows should be calving every year. With an average lactation length of 305 days there shall be:

100%·305 (average number of days milked)/365=83.56% of animals in milk round the year. The percentage of animals in milk round the year will vary according to the average intercalving interval and the days milked, and can be calculated using the formula. An example is shown in Table I.5.

Cows in milk at any given time (%) =
Average days milked/Average calving interval in days · 100.

**Table I.5** Percentage cows in milk round the year according to changing calving intervals and average days milked

| Days milked | Average calving interval in days | | | | |
|---|---|---|---|---|---|
| | 400 | 410 | 420 | 430 | 440 |
| 285 | 71.25 | 69.51 | 67.86 | 66.28 | 64.77 |
| 295 | 73.75 | 71.95 | 70.24 | 68.60 | 67.05 |
| 305 | 76.25 | 74.39 | 72.62 | 70.93 | 69.32 |

Other parameters for estimating the breeding efficiency of a herd are conception rate, calving rate, artificial insemination index and service period.

**Conception rate** shows the percentage of success of artificial insemination. The average conception rate for a month can be calculated using the formula:

$P / TAI \cdot 100$

where: TAI is the total number of artificial inseminations conducted during the second preceding month; P is the number of cows tested positive out of the artificial inseminations conducted during the second preceding month
**Calving rate** is the percentage of calves born out of the total AI done
**AI index** is the average number of AI done for every calf born
**Service period** is the interval between calving and next conception

While the percentage of animals in milk gives the current status of the herd, other measures give the status of the herd at an earlier date. As such it is important that the manager of the farm takes immediate corrective measures when the percentage of cows in milk in a herd drops below the fixed norms. Software packages available for the management of herds have many check features to generate lists of problem animals. This is possible however, only when the information is fed to the computer on a daily basis.

### Mating programme

The semen of only top quality bulls must be used for breeding bull mothers. Only top proven bulls must be used and in their absence high pedigreed bulls shall be employed. The bulls used must be of the desired genetic groups as specified in the breeding policy. A mating programme for individual cows is to be prepared for a year and must be practised. While preparing the mating programme care should be taken to arrange mating between unrelated individuals. In a crossbreeding programme it is required to have a wider genetic base for which a bull to cow ratio of 1:10 is recommended. A bull allotted to a cow may be changed after getting one progeny in order to obtain different genetic combinations.

## Estate management

Management of the farm is one of the most difficult tasks. Many of the routine activities such as milking, cutting and transporting of roughage, dung cleaning, and watering can be mechanised to improve efficiency. The number of workers required for the various activities of a farm varies according to the situation prevalent in the area/country. Training and orientation of the employees and making them aware of their responsibilities are vital to the success of farm management. Frequent interactions with them, considering their views while making major changes in the operations of the farm, giving them information about the economic aspects of the farm and providing them motivation through welfare and other measures will help in efficiency improvement. It is essential to have the optimum labour strength; too many employees will reduce the efficiency as well as increase the running cost; less than the optimum number will result in many works done improperly.

The risk of management error can be reduced by:

- Keeping good records to monitor and evaluate results
- Establishment of realistic goals
- Providing the specific information for scheduling and executing action

# 4. Frozen semen production

*Frozen semen production would deserve a book on its own. This chapter therefore banks on various existing standard publications and treats the issue in a wider sense. It starts with the housing and management of bulls. The training of AI bulls is a central element. Breeding bulls require specific management, feeding and health care, which can be monitored and supported by systematic need-based record-keeping. The design and layout of frozen semen bull stations is discussed before embarking on semen collection, processing, storage and dispatch systems. Manpower requirements both in quantity and quality are discussed based on the Kerala experience.*

# Breeding bulls

## Training and handling of AI bulls

Good bulls maintained well and physically fit to produce semen, are vital for genetic progress. The bull calves that are primarily selected as breeding bulls are separated from adult females by the time they attain sexual maturity. A nose ring of the suitable size is applied to all selected bull calves before they are taken for training. Metals or alloys that do not rust and having sufficient strength are used for making the nose ring. Chains or ropes tied around the neck and attached to the nose ring are used for controlling the bulls. Training in handling and leading must be given to the bulls when they are young. Young bulls that are sexually mature as evidenced by their phenotype and behaviour in the group are taken for training during normal collection days. The young bull during the training period must be handled with care and affection so that it willingly comes to the collection yard and starts mounting on dummies.

Dedicated and trained employees who are caring and understanding will make the management of the bulls efficient and ensure human as well as animal safety. The bull attendant must lead the bull from the side holding the nose always higher than the natural level. As a normal practice, visitors should not be permitted to the bull stud and those who are permitted must wear protective footwear and aprons for aseptic reasons. The bull handler who brings the bull for collection must be responsible for keeping the bull under control and safely away from other persons and bulls in the collection yard. He is also responsible for appropriate mount animal restraint unless a separate handler manages it. Proper training of young bulls is very important to reduce the percentage of bulls culled on account of poor reproduction (not mounting, no thrust and poor semen quality). Selected young bull calves are reared in a group to give them a chance to mount each other and during the play the penis gets separated from the sheath. Young bulls have to get acquainted with the attendants, methods of restraining and handling. During the training period collections are taken once a week. The training lasts generally four to six months during which period a set number of collections are to be made. Any pain or injury during the early days of training will make the young bull reluctant to mount. It is a good practice to bring the young bulls to the collection arena early and keep them there till the semen collection is over. The selection decision is to be taken based on the animal's performance during this period supported by information on mounting behaviour, thrust, semen quality and libido.

## Feeding

Bulls should be fed a nutritionally balanced ration using locally available good quality feed and fodder. One kg hay and 0.5 kg compounded cattle feed per 100 kg live body weight is fed to a breeding bull. Bulls may be fed once or several times daily. Clean and good quality water must be readily available.

## Health care

Bulls must remain in good health and physical condition to achieve optimum growth and maximum production of quality semen. Since testis size is highly heritable (Coulter et. al, 1976), monitoring the scrotal circumference is found to be a good check on the reproductive ability of a bull. Sudheer (2000) observed a significant positive correlation between scrotal circumference and ejaculate volume and sperm output in crossbred bulls. Regular exercise will keep the bulls in good physical fitness and will reduce the incidence of hoof problems. Overgrown hooves must be cut and trimmed. Excessive hair growing around the prepuce has to be cut to avoid matting and contamination to the collected semen. Trained and experienced veterinarians shall examine the bulls regularly for health problems. Annual or half-yearly surveillance testing for the bovine contagious diseases prevalent in the area must be a major component of the health management programme in a Deep Frozen Semen Production Station (DFSPS).

## Culling

Timely disposal of unwanted bulls is often not given due importance in many of the DFSPSs of tropical countries. Culling decisions must be taken only after a comprehensive review of the bull's spermatozoal morphology and semen production ability with the help of proper records. The major reasons for disposal of crossbred bulls in India are reported to be poor semen quality and poor libido (Bhosrekar, 1988), serving ability, sperm motility, spermatozoal output, and post-thawing revival rates (Sudheer et. al, 2001).

Even when the bull is producing good quality semen it is removed from the herd after the production of a certain number of doses to give room for younger bulls. This is often the practice in the case of crossbred bulls to involve a larger number in the programme. Bulls are also removed from the stud when they are very old, aggressive or physically handicapped.

## Record-keeping

Keeping records of pedigree, and all the reproductive and semen production attributes of the bulls are highly essential for semen quality analysis and for taking management and selection decisions. Of late, personal computers with user-friendly software are widely used for data management.

# Deep frozen semen production station (DFSPS)

Deep frozen semen doses are increasingly used for artificial insemination in the developing countries. Production of semen doses of the required genetic groups at cheaper cost are possible by the establishment of frozen semen production stations within the artificial insemination organisation. The DFSPS can be an independent unit or part of a bull mother farm. In either case it should have facilities to:

- Maintain the planned number of adult breeding bulls and young bulls in training
- Produce and store semen doses according to the production plan
- Produce/procure and store liquid nitrogen necessary for the processing and storage of frozen semen

## Location and size

The DFSPS is to be located in a place where natural resources and physical characters are satisfactory. The number of bulls maintained in a Deep Frozen Semen Production Station is related to the requirement of frozen semen. While small stations are easy to manage they would necessarily increase the cost per dose of semen produced. Mathew (1984) recommended bull stations with a capacity of 70 – 80 bulls as optimum for Indian conditions and further stated that managing more than 100 bulls in a station would be difficult. Where the DFSPS is part of a bull mother farm, care has to be taken to keep it as a separate complex. The various buildings in the Station, such as bull sheds, bull exerciser, fodder store, semen processing laboratory, etc. have to be located in a logical sequence to make the work efficient and effortless and always in consideration of the topography. Figure I.13 gives an example of the location of the various buildings.

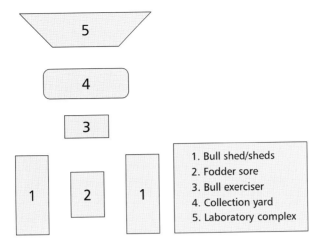

**Figure I.13** Placement of the various structures in a Deep Frozen Semen Production Station

## Infrastructure

### Bull shed

Bulls are housed in clean, well-lit, ventilated buildings or outside with facilities that protect them from weather conditions. Even though cattle tolerate a wide range of ambient temperatures, housing should be designed to protect bulls from extreme heat or cold. It is reported (Bhosrekar, 2003) that the sex libido and semen quality of exotic and crossbred bulls are reduced during summer season in tropical countries. It is therefore necessary that the bulls be housed in ventilated, cool and dry sheds. Bulls housed outdoors should have access to shelter for protection from sun and severe weather. The design of bull sheds would vary depending on geographic location and feeding and management practices followed. Internal surfaces of the sheds must allow for effective cleaning and disinfecting and must not have sharp edges or projections. Provision has to be made for the segregation of sick or injured animals as well as for quarantine. Waterers, ventilator fans, heating and lighting units, fire extinguishers, alarm systems, etc. should be inspected regularly to ensure their proper working. All electrical installations must be inaccessible to cattle. The designs of bull sheds should allow for social interaction between bulls both visually and vocally. The facilities in the DFSPS and management practices should provide safety for bulls and handlers. In general, the performance of the bulls should be the primary indicator of the adequacy of the environment. Keeping bulls in a single or double row shed has the advantages of:

- Less space per bull
- Labour saved on feeding, cleaning and watering

- Better visual and olfactory contacts between bulls
- Easy management
- Cost effective

Attention should be given however, to providing enough exercise to the bulls and to preclude fighting between them.

## Slurry tank

The slurry tank to collect and store the dung and urine preferably underground and covered is ideal. The fodder store should have the capacity to store enough roughage for about six months and shall be located at a place from where the fodder can be transported to the feeding trough with minimum effort. A shed with a capacity to keep around 5 to 10% of the total strength of the bull shed has to be constructed away from the complex to keep the isolated animals and animals in quarantine. The collection yard should ideally be located in between the bull sheds and the semen processing room.

## Semen collection yard

The size of the semen collection yard depends on the number of bulls collected on an average collection day and should be such that the individual bulls and mount animals can be handled freely. Mount animals are dummies for the bulls to mount and generally regular bulls in collection are alternated as mount bulls. The floor must be soft but firm and provide good 'footing' to prevent slipping and subsequent injury to animals.

## Laboratory

The semen processing room can be a rectangular hall with the semen receiving counter at one end and the semen dose storage area at the other end, with attached rooms for change of uniforms, sterilisation and preparation of the artificial vaginas (AVs). In locations where the atmospheric temperature goes above 30°C and where there are problems with dust, the semen-processing laboratory must be dustproof and temperature controlled. Providing shade by planting trees around the semen processing complex, cultivating fodder grass around and black topping the roads are measures that will keep the location dust-free and at a lower ambient temperature. Facilities to accommodate the office of the manager, a technical library, store, liquid nitrogen storage room, meeting room and a pantry are also necessary in the DFSPS.

## Equipment

A Deep Frozen Semen Production Station needs a long list of equipment and machinery, many of which are very expensive and the operation of which needs special skills. The list of equipment and machinery cannot be complete as well as up to date since new requirements and modern equipment are entering the market very fast. However, a comprehensive list of equipment and machinery for a DFSPS is given in Box I.7 indicating the purpose of each of the items.

**Box I.7**  List of equipment required for deep frozen semen laboratory

| Name | Quantity | Notes |
|---|---|---|
| Generator | I | As a stand-by power source. Only if the electricity supply is irregular |
| Hot air oven 75·75·75 cm | I | For sterilisation of glassware |
| Artificial vagina (AV) | 60 | |
| AV Steriliser | I | For sterilisation of AV |
| Water bath with built-in stirrer 50·50·15 cm | I | For keeping neat semen for initial evaluation |
| Microscope with hot and cold stage and self-illumination | I | For semen examination. Preferably phase stage and self- contrast |
| Photoelectric calorimeter | I | For deciding the semen dilution rate. New computer controlled system, e.g. semen analyser, would also be good but expensive |
| Refrigerator | I | For storing the diluent and chemicals |
| Cold handling unit | I | For handling the diluted semen |
| Automatic filling and | I | For packaging semen sealing machine |
| Wide-mouthed liquid | I – 2 | Semen freezing nitrogen refrigerator |
| Precision balance | I | For weighing the chemicals. Preferably electronic |
| Printing machine | I | For printing the identity of the bull on the straw. New generation jet or laser printing machines available |
| Drying cabinet | I | For drying the glassware |
| Double distillation apparatus | I | For producing ion-free distilled water for dilution. All glass type is necessary |
| Straw distribution ramp | I | For spreading the filled straw in the rack |
| Freezing racks | 30 | For keeping the straws for freezing |
| Valsallum forceps | 6 | For withdrawing straws from the container 45 – 50 cm long |
| Freezing grill | I | For keeping the freezing racks in the nitrogen vapour |
| Biological freezer | I | For freezing the straws |
| Desktop computer with the necessary software | I | For inventory of semen, documentation, and as part of the semen processing system |
| Narrow-mouthed semen storage refrigerator | | For storing semen. Quantity depends on capacity and the doses to be stored |

Contd.

| **Box I.7** Contd. | | |
|---|---|---|
| Narrow-mouthed medium size semen storage refrigerator | 10 | For dispatching semen to other stations |
| Liquid nitrogen containers 25 – 50 L capacity | 15 | For transferring LN |
| Goblets | 2000 | For keeping semen straws |
| Small refrigerators | 10 | For dispatching semen to AI units |
| Liquid nitrogen tanker - capacity 5000 L | 1 | For storing liquid nitrogen. An LN plant can be installed where LN can not be purchased from outside sources |

Liquid nitrogen is a critical element of a Deep Frozen Semen Production Station and has to be readily available at all times. Liquid nitrogen is required for semen processing, semen storage at different places (DFSPS, Regional semen supply station and AI units) and for semen dispatch. It may also be remembered that at every point of transfer of liquid nitrogen from one container to another 15 to 25 per cent of the liquid nitrogen escapes as gas. The liquid nitrogen requirement of a 50-bull DFSPS is around 700 L for semen processing and 700 – 800 L for semen storage per year.

# Semen collection

## Collection schedule

A schedule of semen collection must be drawn up for each bull for maintaining optimum sexual function of the bull as well as for maximum efficiency of the DFSPS. Usually bulls are collected at three or four day intervals. Taking double ejaculations each time has become almost a conventional practice. About 80 to 90 collections are taken from a bull in a year. Semen collection must be accomplished maintaining healthy and hygienic practices, safety for bull handlers and meeting the required quantitative and qualitative standards. Collecting semen using an AV is the most commonly adopted practice. The artificial vagina can vary in diameter, length and method of controlling the internal pressure. Electro ejaculation is another method practised when the bulls are physically unable to mount or not trained for collection using AVs and should be employed only after a diligent effort to harvest spermatozoa via the artificial vagina has failed. Ampullar massage is yet another semen collection method but is less consistent in harvesting semen of acceptable quality. The process of semen collection using AVs is described below under preparation of artificial vagina, and bull handling and semen collection.

## Preparation of artificial vagina

There are four parts in an artificial vagina assembly: artificial vagina hose, inner liner, artificial vagina cone and collection vial (Figure I.14).

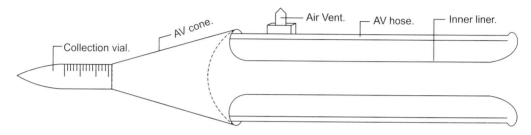

**Figure I.14**   Different parts of an artificial vagina (AV)

The artificial vagina must simulate the normal vagina of a cow in heat with respect to pressure, friction and temperature for successful sperm harvest. It is advantageous to keep a constant AV temperature for a bull rather than changing it every time. As age advances bulls tend to prefer higher artificial vagina temperature. The AV is filled half to three-fourths full with water having a temperature of 40° to 60°C and then air is blown in through the valve to provide the required pressure. Pressing the AV hose (if made of flexible rubber) while taking collection can provide additional pressure if necessary. A sterile and non-spermicidal lubricant such as vaseline or vaginal jelly applied to the upper one-third of the artificial vagina liner will improve the response of the bull and minimize penile abrasions. While young bulls prefer smooth liners slightly older bulls prefer a roughened surface. The artificial vagina is designed in such a way that the semen drains into a collection vial through the cone. Quantity and quality of the semen are affected if the semen is ejaculated into the artificial vagina hose. The AV cone and collection vial must have a temperature of 35°C. An insulated jacket is fitted to cover the collection vial and the cone when the atmospheric temperature is above 40°C and below 25°C for protecting the semen from sudden change in temperature. After the collection is over, the vial is removed, labelled properly and dispatched for processing.

On average 2.5 artificial vaginas are necessary per bull collected when double ejaculations are taken. The required number of AVs must be kept ready without lubrication and attachment of the collection vial, in a warm cabinet maintained at 45°C. This can be done the previous evening when the semen collection starts in the morning.

Artificial vaginas are to be cleaned soon after use using a dairy cleaner. The AV liners and cones after cleaning must be immersed in 70% alcohol solution for five minutes for sterilisation and afterwards dried in a clean cabinet by hanging them. The artificial vagina hose is sterilised in boiling water. All the glass articles are sterilized using standard procedures.

## Bull handling and semen collection

To prevent potential transmission of diseases during semen collection, the hindquarters of the mount animal must be thoroughly disinfected between successive mountings. A clean apron is tied to the belly of the mounting bull to prevent contamination of semen collected. A separate artificial vagina or AV liner must be used for each ejaculate. The semen collector determines when the bull is properly prepared, when the semen collection can be safely taken and performs collection of semen from individual bulls. He must ensure that procedures are hygienic and that the collection vial containing the semen is correctly identified with the number of the bull.

Proper sexual stimulation is necessary for getting good quality semen in optimum quantity. The presence of other animals in this area and various visual, olfactory, and auditory stimuli, sexually arouse the bull and will increase the semen output (Hafs, 1972, Burman, 2003). The various methods employed for proper stimulation are:

- False mounting or half mounting
- Taking the bull around the collection yard
- Moving the mount bull forward and backward
- Changing the mount bull
- Changing the location of collection
- Bringing another bull near the mount bull

When the bull is sexually aroused the penis will be erect and will want to mount other bulls and/or a mount animal. Depending upon the libido of the bull and the frequency of collection, the stimulation may be accomplished in a matter of minutes or it may take a longer time. To prevent contamination of the penis and possible injuries to it, the semen collector should be at the bull's side to hold the sheath aside avoiding contact of the penis with the rear quarters of the mount animal.

The semen collector shall have protective clothes and boots, should change his apron after each collection and should wash and dry his hands after every collection or use disposable hand gloves. He should get the sterilised ready-to-use artificial vagina from the artificial vagina room and deliver the collected semen after affixing the identification mark of the bull on the vial to the processing laboratory through a window without entering the laboratory. He should work with the bull throughout the preparation procedure and determine

*The breeding bulls are alternated as dummy for collection of semen using artificial vagina.*

the optimum time for semen collection. During the stimulation period he should stand quietly a little away from the bull and watch. Mounting should be from straight back of the mount bull. On mounting the semen collector should reach the side of the bull, take a balanced stand in the direction of the bull and leaning on the shoulder of the bull, grasp the sheath and direct the penis to the side. By this time the bull will clasp the mount bull and move the penis in all directions in search of vulva. If the stimulus is found satisfactory offer the front end of the artificial vagina to the glans penis and hold it firmly without pushing it backwards for the bull to give a strong thrust. Ejaculation takes place during and immediately after the thrust. The artificial vagina is withdrawn slowly as the bull dismounts. Generally the semen collection is performed immediately after the false mounting regime is completed.

There are two internationally accepted semen collection regimes, single ejaculation regime and double ejaculation regime. In the single ejaculation regime the bull is brought for collection three times in a week and each time a single ejaculate is collected. Two collections are taken each time the bull is brought for collection and collections are taken once in a week in the latter case.

# Semen processing

## Examination

The semen received from the collection arena is to be kept soon in a water bath at 27–32°C until the preliminary dilution is completed. Individual bull's semen received in the laboratory has to be evaluated for maintaining the quality. Storing fresh semen without dilution for more than 30 minutes increased the number of abnormal spermatozoa (Rajamannan et al., 1971). On physical examination the semen should have uniform appearance and should be free from contaminants, e.g. hair, dirt, dung, pus, urine, blood, vaseline, etc. The volume of the semen can be noted as the collection vials are calibrated. Too low volumes can be a result of frequent collections, genetic make-up of the bull and/or insufficient thrust while taking collections and will affect only the number of doses that can be produced if the semen is otherwise good. Motility of the semen is estimated by placing a drop of semen with a cover slip on and examining it under a microscope (preferably a phase contrast) with a stage warmed up to 37 – 40°C. A reasonable assessment of the fertility of the bull is possible based on the post-thaw motility of the semen. Samples having less than 50% progressively motile sperms may not be accepted.

Proper utilisation of semen without wastage depends on a reliable estimate of the concentration of semen. There are many methods to estimate the concentration; however photometers are found to be good enough value for the money for small stations. More expensive and modern equipment such as semen analyser, electronic counting machines etc. are now available in the market. It is up to the Deep Frozen Semen Production Station to select equipment that suits its requirement and budget.

## Dilution

Diluents are used to increase the volume of semen and liveability of spermatozoa. It should provide protective factors, metabolites and buffer salts to enhance the lifespan and activity of the sperms and should not be harmful to the sperms. Tris-glycerol diluents, glycerol egg yolk citrate, milk glycerol diluents, and other commercial diluents, are commonly used for diluting bull semen. In addition sugar (fructose) for additional energy source, enzymes (amylase or beta glucuronidase) for liveability and antimicrobial agents (sulphonamides and antibiotics) are also added to the diluents. Semen extenders containing soybean-derived components are commercially available (Biocephos -plus marketed by M/s IMV France) as replacement for egg yolk citrate (Hinsch et al. 1997; Haard, 1997). The non-return rates obtained were similar from semen doses processed with egg yolk tree diluents and egg yolk diluents when the sperm per dose were in the range of 15 million (Nehring and Rothe, 2003).

Dilution of semen is done in two stages, the first soon after accepting the semen at normal temperature and the second with glycerol-containing fraction of the diluents after slowly cooling the diluted semen to 5°C. After dilution and just before freezing the diluted semen is stored at 4 – 5°C for a few hours to obtain better post-thaw motility and fertility. This pre-freeze storage time is referred to as 'equilibration time'. The equilibration time is generally between 2 and 3 hours.

## Packaging

There are various schools of thought regarding the number of spermatozoa to be packaged in a dose of semen. Generally it can be stated that the total number of spermatozoa to be packaged in a dose of semen ranges between 10 to 30 million with a view to not less than 5 million live spermatozoa. However, the total number of spermatozoa per dose as well as the number of live sperms were gradually brought down over the years. It was observed that the post-thaw quality of spermatozoa increased with decrease in package volume; however, sperm quality improvement was relatively small (Devanand, et al., 2003). The French mini-straw system wherein 0.25 ml of diluted semen is packaged in a polyvinyl chloride straw 13.5 cm long and sealed, is the most commonly used method. Another system less popular but still practised is the mini-tub procedure where the length of the straw is 6.5 cm and the capacity is 0.26 ml. Advantages like ease in handling and transport, smaller volume, negligible sperm loss, higher recovery rates of sperm on thawing and improved conception rate make the 0.25 ml packaging very popular. The identity of the bull, breed and collection date are printed on the pellet before it is filled. Filling and sealing machines, both automatic and manual, manufactured by different firms are used for packaging semen.

## Freezing semen

The semen doses arranged in horizontal racks are gradually cooled to the deep frozen stage in static nitrogen vapour by keeping them over a grill kept in a wide mouthed nitrogen container. The grill should be at least 30 cm above the liquid nitrogen level. After 10 minutes, by which time the temperature of the semen dose reaches –40°C, the doses are collected from the rack, put in a pre-cooled goblet and plunged into the liquid nitrogen for storage. The rate of freezing can be adjusted by varying the height above the liquid nitrogen. Programmable freezers, which are now commercially available, have facilities to precisely regulate the rate of freezing. Studies from many laboratories however, indicate that freezing semen in static nitrogen vapour or using a programmable freezer produces similar results (Devanand, et al., 2003). Many of the DFSPS are gradually switching over to biofreezers.

# Frozen semen storage

The experience and training of the technicians in correct semen handling procedures become highly critical to the fertilising quality of the semen doses once they leave the laboratory. Once the semen doses are kept in liquid nitrogen, one should be careful not to break the cold chain. It was shown that by changing the temperature from –79 to –196°C and vice versa, the fertilizing power of the sperms reduced (Kalev and Zagorzki, 1968). Exposure of semen to room temperature is likely to result in intra- or extracellular ice crystal reformation, thereby damaging the cell membrane. The storing of semen in refrigerators should be done in such a way that its recovery is easy, locating the goblets is simple and space utilisation optimum.

There are different types of goblets (the cylindrical container made of plastic or aluminium) to suit the different makes and sizes of refrigerators. It is recommended not to fill the goblet fully with the straws in order to allow free flow of liquid nitrogen into the goblet, to allow speedy handling and to prevent breaking of the straws while withdrawing. The goblets are put into a perforated cylindrical tube with a handle, called a canister. There are different types of canisters that can hold one or more layers of goblets. It is necessary to have a proper recording system about the location in which the semen of a particular bull, a particular batch, etc. is stored. Since the various places where the canisters are attached in the refrigerator are numbered, the storage place for a particular bull's semen can be identified with these numbers. The inside of bigger refrigerators is divided into compartments, which can be identified as per the specifications given. The storage capacity of the DFSPS shall be around half its yearly production capacity.

Frozen semen can be kept for long periods without affecting its fertility if properly stored in liquid nitrogen. At least half the length of straws in storage should be below the level of liquid nitrogen. Further lowering of liquid nitrogen level will lead to decline in post-thaw motility and fertility of semen. Tight packing of semen in goblets causes physical damage to straws at the time of packing as well as withdrawal. Moreover, tight packing will not allow entry of

*The number of frozen semen doses required in the AI unit is transferred to small refrigerators at the regional semen supply station.*

liquid nitrogen inside the goblet, which will endanger the safety of the semen. The storage of straws has to be regulated so that semen can be withdrawn batch-wise and bull-wise. Canisters and goblets in which the semen is stored are identified and tagged properly in order to avoid possible mistakes at the time of dispatch to AI units. The position of each goblet is to be recorded and identified. It should also be ensured that while distributing, semen stored in one goblet is fully distributed, before semen from another goblet is taken for distribution. When the stock of semen is low, the goblets must be positioned in the lower layers of the canisters. In each goblet a thick strip of paper in which the details of semen are recorded should be inserted and entries made in the strip whenever semen from that goblet is transferred for field use. The storage of semen should be planned in such a way that:

- Transfer of straws between goblets is reduced to the minimum
- Semen of bulls expected to be in high demand and fast moving is easily accessible
- Straws meant for long storage are always at the bottom layer of the canister

# Frozen semen dispatch

The Deep Frozen Semen Production Station dispatches semen doses either wholesale or retail as the case may be, depending on the arrangement with the artificial insemination units. When a large number of artificial insemination units are catered by the DFSPS generally the dispatch of semen to the artificial insemination units is carried out by a second level unit, a regional semen supply station. In this case the semen doses of the required genetic groups and numbers are dispatched from the DFSPS in big refrigerators to the regional station.

Handling of semen during dispatch can be minimised once the goblets in both the stations are interchangeable and when each goblet is identified and location details are known. Transferring semen goblets from one refrigerator to another should be done fast, keeping the two refrigerators adjacent and using a long forceps. When semen doses are to be counted and or transferred between goblets, it should always be done under liquid nitrogen. A wide-mouthed liquid nitrogen container or a styrofoam box filled with liquid nitrogen can be used.

# Manpower requirement

Trained and sincere human resource is very vital for the management of a DFSPS. The manpower necessary in a DFSPS can be grouped under professional managers, technical assistants and labourers.

A graduate in animal husbandry and/or veterinary sciences with managerial skills shall be a suitable person as the manager of the DFSPS. Assistant mangers shall support him; the number will depend on the size of the unit. In the normal course two professionals—at least one a veterinarian—shall be put in charge of the management of bulls and laboratory. If the unit is small the manager himself can handle one of the functions.

Technical assistants shall be assigned jobs such as semen collection, hoof care, vaccinations and first aid, filling and sealing of semen straws, printing, dilution etc. The technical assistants should be given sufficient training in their respective jobs and routinely retrained.

For the upkeep of the bulls, handling them during semen collection and exercise, cleaning and sterilisation of the laboratory and equipment, night watch etc., unskilled workers are employed. The numbers necessary for carrying out the works satisfactorily depend on the strength of the Deep Frozen Semen Production Station.

# 5. Management of artificial insemination service

*There are various systems, procedures and methods to run an artificial insemination service. However, the information on practical examples is scarce. This chapter discusses a system as practised in India from the description of an artificial insemination unit to the qualification of the artificial insemination technician. Procurement and dispatch of the semen as well as the recording and monitoring system are important elements of the service. The material and the logistics to provide liquid nitrogen are discussed and various options are illustrated.*

# Introduction

In many of the developing countries management of an efficient artificial insemination service is not always easy. A study on coverage of artificial insemination in India (NDDB, 2002) revealed that in many states there has been no appreciable increase in coverage of artificial insemination since 1990. Major reasons pointed out are:

- Shortage of breeding bulls for semen production
- Shortage of skilled technicians and extension workers
- Lack of conviction of farmers about the success of artificial insemination
- Unsatisfactory quality of the product and services
- Slow adoption rate of artificial insemination by livestock owners

It is necessary to develop area-specific programmes based on the requirements, farmer needs and infrastructure facilities available. Some of the reasons pointed out for stagnation in the artificial insemination programme in India are worth considering in the planning stage. The artificial insemination service delivery system discussed below starts with the artificial insemination unit and ends in the semen production station. An example of an artificial insemination organisation is shown in Figure I.15.

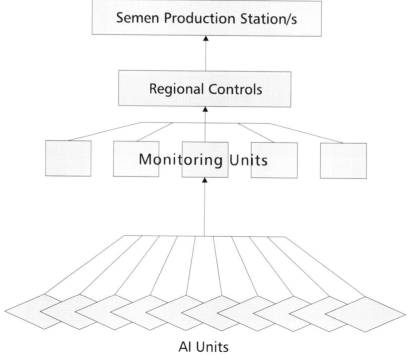

**Figure I.15**  Functional levels of an artificial insemination organisation

*AI using frozen semen is widely used for impregnation.*

# Artificial insemination unit

## Type of artificial insemination unit

An artificial insemination (AI) unit is defined as the primary unit for providing artificial insemination service in a given area. A person trained for carrying out artificial insemination, its follow-up and for carrying out extension services can man an artificial insemination unit. The artificial insemination service can be stationary or mobile. In stationary units cows are brought to the place where the technician is stationed whereas the technician reaches the cow at its abode in a mobile unit. As the cows will be in a better physiological condition at their homestead the conception rate is better in a doorstep artificial insemination service. The milk production loss and risk of dragging the cow through busy townships are also avoided. Good transport and communication facilities must be available for the operation of a mobile artificial insemination programme. However, it will have a lower intensity of animal coverage than a stationary unit due to reasons such as extensive area to be covered, less intense farmer contact and long-distance travel.

The number of artificial insemination units, technicians, infrastructure and equipment required for covering a given area with artificial insemination service is calculated based on the type of service provided. The number of animals covered will increase if the technician is moving around in motorised vehicle and if the intensity of cattle/buffalo population is high. A comparison between mobile and stationary systems of artificial insemination service delivery in India based on situations prevailing in 2003 showed that mobile service is considerably cheaper than stationary (Chacko, 2003). The readers are invited to use Exercise 4 given in the toolbox of the attached CD to make similar comparisons with varying figures and assumptions.

*Exercise 4.    Cost comparison between mobile and stationary AI services*

It is recommended to begin the artificial insemination service to cover a limited number of animals in an area that can be effectively controlled. The programme can be expanded making changes if required, based on the experience gained.

## Artificial insemination technician

The artificial insemination technician is either a paid employee of the organisation or a self-employed person. The experience in India revealed that artificial insemination service rendered by paid employees is less efficient than that provided by the self-employed person with regard to quality as well as cost recovery. Recently many states in India have started switching over to private artificial insemination service without much subsidy from the government. A study conducted in India (Ahuja et al., 2000) revealed that free/ subsidised artificial insemination services provided by the governments were in effect not free or subsidised while reaching the farmer and were of inferior quality.

# Monitoring and control
# of artificial insemination units

A good monitoring system is to be designed and implemented from the beginning of the programme. The number of artificial insemination units that can be effectively monitored by one unit will depend on the local situation, the type of artificial insemination service and the type of technician. Monitoring and control mechanism of the artificial insemination organisation must also be binding for the private technicians in the interest of the public as well as the state. As there would be less administrative control over the private technicians, the number of private artificial insemination units monitored by one station could be more than that of paid employee units.

Timely collection of data is a major activity of the monitoring unit. Formats to collect data, periodicity of collection and checks and controls on the data are to be decided by the Regional Semen Supply station and communicated to all concerned. An effective farmer contact system is essential for success of the artificial insemination programme in developing countries. The monitoring unit must take up farmer awareness creation programmes and provide support for reproductive health of the cattle, castration of unapproved bulls in the area and other activities necessary for intensification of the artificial insemination programme.

A systematic management of information to assess the impact of the programme with respect to cost recovery and planning, animals covered, success rate, efficiency, etc. is necessary from the very beginning. Good software packages are available for data management and efficient monitoring.

# Regional semen supply station

## Purpose

The regional semen supply station carries out the regular supply of frozen semen and liquid nitrogen to all the artificial insemination units under its jurisdiction. Around 300 to 400 AI units can be controlled by one regional semen supply station. The responsibilities of the regional semen supply station are:

- Procuring the required number of frozen semen doses

- Supply of frozen semen as per the breeding policy and requirement of AI unit
- Supply and maintenance of artificial insemination equipment and stationery
- Production/procurement and supply of liquid nitrogen
- Co-ordination between the monitoring units within a region
- Data collection, analysis and feedback

Semen distribution requires careful planning to ensure that sufficient doses of frozen semen of the required breed and bull are available at the AI unit. A schedule for the supply of frozen semen, liquid nitrogen and other consumables to the artificial insemination units has to be drawn up in advance. All the units must know the date and time of supply well in advance so that arrangements to receive the consignment can be made. The quantity of semen doses needed in the unit according to breed must be intimated to the regional semen supply station in advance. One possibility is to give an indent at the time of the supply so that the indented items can be brought during the next supply.

The regional semen supply station must have facilities to procure, store and dispatch frozen semen to the artificial insemination units under its control. Frozen semen is either purchased from outside sources or produced within the AI organisation. It would be easier to purchase frozen semen, especially when the requirement is small. But when the requirement is more and/or when the semen of the required genetic groups of bulls is not readily available, production of frozen semen within the organisation is the option.

## Operational aspects

### Semen procurement

The regional semen supply station must stock a sufficient quantity of the requisite types of semen. Before taking delivery from the production station, the semen must be checked at random for motility. Only semen with more than 40% post-thaw motility must be taken from the production station.

At the end of each year the stock of frozen semen kept in each refrigerator must be verified goblet-wise. The stock balance as per records, the actual stock and bull-wise stock must tally. When semen from one goblet is issued in full it must be verified whether the total number issued tallies with the number received or purchased.

### Semen distribution

In the normal course the semen doses procured by a Regional semen supply station should be distributed within 3 months of receipt. The following procedure is recommended while transferring semen for the distribution to the artificial insemination units:

- Use a thermo–cool box filled with liquid nitrogen
- Transfer the semen doses from one goblet to the other without delay
- Cool the end of forceps by dipping in liquid nitrogen before the semen straws are touched
- Pool semen doses of the same bull after each transfer to save liquid nitrogen and space

Semen doses from the regional semen supply station are transported to the artificial insemination unit in liquid nitrogen filled refrigerators. After reaching the unit first fill the refrigerator of the unit with liquid nitrogen and only then transfer the semen doses. The canister containing the semen doses must be transferred to the refrigerator of the unit within the least possible time. Before closing the container with a neck plug ensure that the neck plug is dry, no water droplets are sticking to the plug and the neck is not filled with liquid nitrogen. Further, the neck plug must not be inserted immediately after the container is filled with liquid nitrogen. Liquid vapour should be allowed to escape freely, otherwise neck plug seizure with the neck tube could damage the refrigerator. Officers in charge of Regional semen supply stations must periodically visit the AI units and monitoring units to discuss and sort out problems. The details of semen and other consumables supplied should be entered in the registers maintained at the AI unit and acknowledgement obtained form the technician.

The truck used for supply of liquid nitrogen and frozen semen to the artificial insemination units must be driven at a moderate speed, must have good antivibration mounts, tight fitting rubber pads or compact crates to avoid damage to the containers and spoilage of semen doses and liquid nitrogen. There must be a 20 mm-thick rubber mat on the floor of the vehicle or inside the refrigerator cage.

## Genetic control

Farmers seldom choose individual bulls and or its breed. In low-income dairy production with small farmer crop livestock mixed farming systems, it is difficult to provide semen of the bull of choice to the farmers. In such a situation it is recommended that germplasm available with the artificial insemination organisation be distributed in the targeted area in a more or less uniform manner and care taken not to increase the level of inbreeding. This is achieved by grouping the bulls into different families with related bulls always placed in one family. The semen of bulls of one family is distributed to one area for around three years and thereafter changed to the next area. This enables prevention of mating between the sire and the daughter. The principles adopted for a bull allotment programme in Kerala, India are described in Box I.8.

**Box I.8**   Principles adopted for bull allotment programme, Kerala, India

---

The state is divided into three zones based on the administrative boundaries called districts.

The related bulls used in artificial insemination are grouped into three families called bull family.

New bulls inducted into the programme, not related to the old bulls are allotted to one of the groups, balancing the group size.

The number of bulls in each bull family is balanced to avoid shortage of frozen semen in a particular family.

Young bulls are allotted to the family of old bulls to which they are related.

Frozen semen from bulls in one family is distributed in one zone for a period of three years, then to the next zone and so on.

In this manner bulls of one family will be going to the same zone after 9 years, by which time the related animals would have disappeared.

The breeding and research department of the artificial insemination organisation controls the bull allotment programme.

---

# Liquid nitrogen

## Liquid nitrogen for semen preservation

For successful implementation of an artificial insemination programme, an uninterrupted supply of liquid nitrogen is very important. Availability of liquid nitrogen and a well-run transport system are essential to the success of an artificial insemination programme. By following an effective plan the supply of liquid nitrogen to a number of AI units in a determined area can be done, minimising the transport charges and time. The liquid nitrogen can be:

- Produced by the artificial insemination organisation
- Procured from bulk producers and suppliers of liquid gases
- Combination of these two possibilities

In areas where liquid nitrogen is available from industrial sources, purchase from such sources is the best choice. Uninterrupted availability and efficient transport are to be ensured however. Producing liquid nitrogen by the AI organisation is necessary when outside supplies are not dependable/available. Many AI organisations in developing countries had to invest in liquid nitrogen plants when industrial supplies were neither available nor dependable.

**Use of liquid nitrogen**

In an artificial insemination organisation liquid nitrogen is required for:
- Freezing of semen doses in the semen production stations
- Frozen semen storage at the production station and regional station
- Transferring frozen semen from one refrigerator to another
- Transport of frozen semen from production station to regional stations and AI units
- Cleaning and testing of refrigerators
- Storage frozen semen at the artificial insemination units

A large quantity of liquid nitrogen (10 to 20% of the total) is lost by evaporation at different occasions of transfer. Therefore, while estimating the requirement the anticipated evaporation losses must also be taken into consideration. The estimated annual requirement of liquid nitrogen for the various activities inclusive of evaporation losses is shown in Box I.9.

**Box I.9**  Estimated annual requirement of liquid nitrogen (LN) in an AI organisation

| Purpose | Units | Quantity of LN/y |
|---|---|---|
| Freezing of semen | 1000 doses | 40 L |
| Semen storage at production and distribution stations | 1000 doses | 15 L |
| Transferring of semen from one refrigerator to another | 1000 doses | 10 L |
| Semen transport; production station to regional station | 1000 doses | 5 L |
| Semen transport from regional station to AI units | 1000 doses | 25 L |
| Cleaning and testing refrigerators (capacity 10 to 35 L) | Per vessel | 20 L |
| Storing frozen semen at the AI unit | AI unit | 100 L for stationary & 200 L for mobile |

## Safety aspects

Though an inert gas, nitrogen is harmful in the liquid state given to its extremely low temperature ($-196°C$). Accidental contact of the skin or the eyes with liquid nitrogen or the cold gas may cause injury similar to burns. In case of accidental contact with liquid nitrogen, wash the area with cold water and apply a cold compressor bandage.

On evaporation liquid nitrogen produces about 700 times its volume as gaseous nitrogen, which naturally reduces the oxygen content of the surrounding air. Working with liquid nitrogen in a confined area, without adequate ventilation, can cause asphyxiation. One or two complete breaths of nitrogen vapour are enough to cause unconsciousness. It would be dangerous to bow the head into the liquid nitrogen refrigerator to identify semen straws.

Overflow, splashing and violent boiling of liquid nitrogen must be avoided. Overflow of liquid nitrogen can damage the sealing material, resulting in vacuum loss and thereby damaging the refrigerator.

## Liquid nitrogen refrigerators and transport tankers

There are several types and models of liquid nitrogen refrigerators available in the market. Details of some of the commonly used liquid nitrogen refrigerators manufactured in India are shown in Box 1.10.

**Box 1.10** Details of commonly used liquid nitrogen refrigerators in India

| Particulars | Type of refrigerators | | | | |
|---|---|---|---|---|---|
| | BA7 | BA11 | BA20 | BA35 | BA3 |
| Capacity (L) | 7.00 | 11.00 | 20.00 | 35.00 | 3.00 |
| Full weight (kg) | 12.45 | 16.80 | 29.10 | 42.34 | 7.00 |
| Static evaporation loss/day (L) | 0.135 | 0.14 | 0.118 | 0.118 | 0.15 |
| Static holding time (days) | 56.00 | 80.00 | 175.00 | 281.00 | 25.00 |
| Semen storing capacity (0.5 ml straws at single level) | 720.00 | 720.00 | 720.00 | 720.00 | 720.00 |
| Refilling interval (days) | 22.00 | 28.00 | 63.00 | 99.00 | 10.00 |
| Refilling required per year | 16.60 | 13.00 | 5.80 | 3.70 | 36.00 |
| LN requirement (Ly$^{-1}$) | 98.60 | 101.90 | 86.20 | 86.40 | 54.00 |

### Care and maintenance of refrigerators

The refrigerators are to be serviced whenever they go dry due to one reason or another. It is recommended to clean and sterilise the refrigerators used in the artificial insemination units once in two years. Checking the depth of liquid nitrogen in a refrigerator is done to ascertain whether it has the required level for safe storage of semen doses. If the mouth of the refrigerator is big enough visual examination is possible but in narrow-necked containers a graduated slender solid stick of low thermal conductivity is used to measure the level of liquid nitrogen. Dip the stick into the refrigerator until it reaches the bottom, keep it undisturbed for 5 to 10 seconds, withdraw and wave in the air. Frost will appear in the wetted section and the length of the frosted section denotes the level of liquid nitrogen in the refrigerator.

## Safety to the refrigerator

Welding, piercing the wall and plugging it tightly while holding liquid nitrogen must be avoided as these will cause permanent damage to the vessel. For transport, a wooden/iron crate or similar protective device must be used. Rough handing or tipping on the side will lead to choking of the inside wall and render the refrigerator useless with no apparent external damage noticed. Charging of a dry refrigerator with liquefied gas should be slow. Otherwise the longevity of the container may be adversely affected and also, due to boiling and possible splashing of liquid nitrogen, personnel engaged in the work may sustain low thermal injury. It is not advisable to empty liquid nitrogen refrigerators completely if they are meant for regular use.

## Liquid nitrogen tankers

Different types of tankers are used for transporting liquid nitrogen. The tankers should be managed properly as per manufacturer's instruction to get maximum life. Copper pipes with arm flex casing are used for delivery of liquid nitrogen from the tankers. Do not fill liquid nitrogen more than the maximum capacity of the vessel. Always keep the overflow valve open while filling. Liquid nitrogen to the tune of around 5% tank capacity must be retained in the vessel to keep it cool in order to reduce the evaporation loss and to increase the life of the tanker. The valve should be opened slowly when there is a large difference in pressure on either side of the valve. At the time of initial filling, the liquid valves must be turned alternately clockwise and anticlockwise through a quarter turn to ensure smooth working of valve the spindle. If liquid nitrogen is trapped between the two valves, it will evaporate and consequently high pressure will be built in the line and the tank may burst. Withdrawal must be done under gravity keeping the vent valve open, holding the container straight and filling only up to the bottom of the neck tube. Liquid nitrogen filling and the withdrawal operation must be performed at a pressure as low as possible to avoid splashing losses. If liquid nitrogen is not coming out even after the valve is opened, there may be ice formation on the backside of the valve. Pour normal water continuously over the valve until the ice melts before the valve is operated. Force must not be applied to open or close the valves.

Periodic checking and preventive maintenance are required for trouble-free operation of the tankers. All the pipelines and valves must be checked once a year for leakage by the soap bubble test. Recalibration of the pressure gauge has to be done at regular intervals. The pressure building coil, valves and piping must be cleaned periodically to prevent frosting, especially during the winter and rainy seasons. New components and fittings must be cleaned, degreased and dried properly before installation/use. The liquid nitrogen tanker must be fully drained, purged and dried with dry nitrogen before any maintenance work is undertaken.

# 6. Field performance recording

*Field performance recording is the basis for successful breeding, involving farmers with a small number of animals. Under tropical conditions the organisation and management of a field performance recording system are faced with a number of obstacles, normally not encountered in temperate areas. Field performance recording systems are the platforms for conducting progeny testing of bulls for the artificial insemination services of the respective areas. This chapter banks on the experience of the field performance programme as conducted by the Kerala Livestock Development Board and offers a 'hands-on approach' for the entire range of elements and activities essential for a successful field performance recording programme.*

# Introduction

Animal recording represents the base for the development of livestock production at the national and farm level. Animal recording, especially recording performance of animals reared by farmers is, still unknown in many developing countries however. Field performance recording involves collection of information on the animals belonging to the farmers, and will depend on the necessity, facilities for evaluation, application of the results and finance. Information on growth, reproduction and milk yield are primary, with milk yield the most important. Regular measurements of the milk yield of cows will:

- Provide information for correct feeding and management
- Help in genetic selection decisions
- Assist conservation and development of local breeds
- Provide records for bull mother recruitment
- Generate data for progeny testing and breed comparison
- Make available data for planning of cattle production activities

Monitoring the growth of animals, especially growing stock, is possible only with information on live body weight. Information on breeding provides valuable tools for the management of reproduction. A detailed review of the available data is the first thing to be done before launching a field performance recording programme. Detailed planning and correct and firm establishment of the programme, help to achieve success in the milk-recording scheme.

The animal must be identified and registered and identity regulations, if any, in the country should be abided by. Recording methodologies can be decided by the organisations carrying out the recording; it is necessary however, to ensure satisfactory supervision. A group of animals kept for the same purpose and at the same location shall be regarded as a herd and the whole herd must be recorded to consider the records official. Details of the service sire and the served animal should be recorded at the time of service and that in turn will be the parentage of the progeny. Identity, sex and date of birth of a newborn calf, shall be recorded as soon after parturition of the dam as possible.

# Animal identification

An identity, unique to each animal, visible and complying with legislative requirements, must be the first thing done in animal recording. According to the International Council for Animal Recording (ICAR) guidelines (ICAR, 1995), the identity number may be attached to the animal by a tag, tattoo, sketch, photo, brand or electronic device. Animals that lose their identity device must, as far as possible, be reidentified with their original number. There must be some external marking to indicate an animal identified using an implanted electronic device. ICAR requires an identity number with a maximum of 12 digits (including check digits). A three-digit numeric code for data transfer and storage, and alpha country code shall be added to identify the country of origin.

Identification is possible in many ways:

1. Giving a name to the animal—not practical when large numbers are involved;
2. Allotting a number—the most commonly used method globally. The number may be:

- Branded on the skin of the animal
- Put on a chain and tied around the neck of the animal
- Tattooed on the inner ear lobe of the animal
- Carved on the horn of the animal
- Tagged on the ear of the animal
- Electronic device implanted on the body of the animal or put into the stomach

Of the various animal numbering procedures, the most commonly employed method is ear tagging. It has many advantages:
- Permanent nature
- Clarity of numbers
- Easy to plan and implement
- Less painful compared to branding or tattooing.

Ear tags are made either of metal or high quality polyurethane material. Earlier metal ear tags were the most popular but with the advent of good-quality polyurethane, plastic tags have become the most widely used. The electronic identification system is being adopted in developing countries and its use fast increasing.

Polyurethane tags are lighter in weight, pliable, bigger in size to make the numbers readable from a distance and cheaper.

The plastic ear tag is composed of two elements:

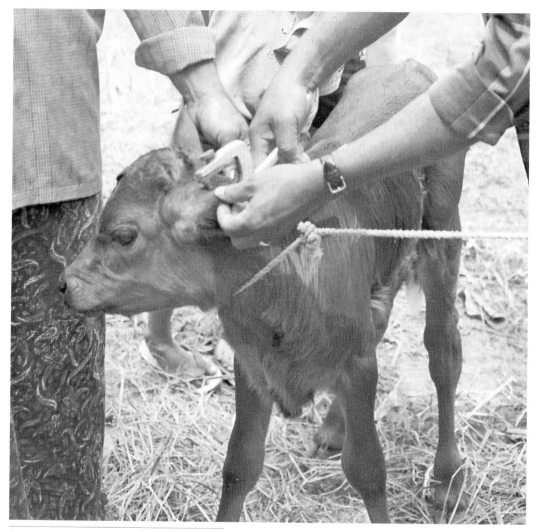

*Animal identification is the first step in any breed improvement programme.*

A front plate, about 5 cm square. The numbers are engraved and painted on this plate. In cheaper ones the numbers are only painted and last less time than the engraved ones;

A rear plate, which, in new generation plastic tags is a button with a nail. The nail pierces the ear lobe and gets locked with the front plate.

Applicators suitable for the different makes of ear tag are supplied by the firms manufacturing the tags. Plastic tags that can be applied in one stroke are more popular in the developing countries, because they are cheap, easy to apply, visible from a long distance and durable.

# Recording milk and milk constituents

## Lactation

Lactation is the synonym for the act of milk secretion. Different orders (or numbers) of lactation can be distinguished, viz. first, second, etc. A lactation period commences on the day the animal gives birth or in the absence of a birth date, the best estimate of the day the animal commenced milk production. As per the definitions given by the concerned organisation it is considered to end:

- The day the animal ceases to give milk (goes dry) or
- The day the animal is milked only once a day or
- The day the animal gives less than the 'minimum' quantity of milk.

The milk produced during the first 305 or less days is denoted as standard lactation. In certain countries 300 or less days is reckoned as the standard lactation. For medium- to low-yielding herds the limit to include short lactations for calculation of herd performance shall be fixed at 210 days. For comparison between cows, standard lactations are considered. Theoretically, different lengths of lactations could be used as standard. By limiting the period of lactation to 305 days, cows with regular reproduction are given their due bonus. If cows should produce a calf every year and if they are dry 60 days before the next calving, 305 days for lactation length is an ideal standard for comparison. Another method, which incorporates the reproductive performance of cows into production estimations, is the calculation of the average yield per day of intercalving period.

Through repeated measurements of the daily milk yield of individual cows and plotting them, the shape of the lactation curve can be established. The lactation curve is characterised by an increase in daily milk production from date of calving to peak milk yield day, a period of stable production, followed by a decline until the cow becomes dry. See Figure I.16 for an example of normal lactation curve.

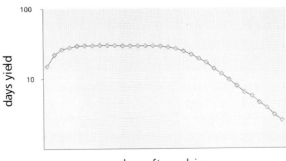

**Figure I.16**   Lactation curve

An animal reaching a high peak milk yield by around 45 days after calving, maintaining the level of milk production for five months (persistency) and gradually decreasing the yield is considered to have produced a good lactation. The major factors that influence the lactation curve are breed of the animal, feeding and management, reproductive status, age, season, health and lactation number.

## Frequency of milk recording

Milk production can be measured by volume or weight. The milk recorded in volume can be converted into weight by multiplying it with the specific gravity of milk. If the specific gravity of milk is 1.03, 1 litre of milk weighs 1.03 kg. The quantity and percentage of milk constituents (butter fat, protein, solids not fat) are estimated based on laboratory tests on the samples collected on the recording date or dates specified for the purpose. The total lactation yield is obtained by recording the milk yield on all days until the end of the lactation and taking the sum. However, in field performance recording programmes where each recording has a cost attached, it is necessary to reduce recordings to the minimum number required for the level of accuracy needed for the purpose for which recording is done. Box I.11 gives the different frequencies for milk recording and their uses.

**Box I.11**  Recording frequency and the probable use of data collected

| Recording frequency | Most commonly used | |
|---|---|---|
| | In | For |
| Daily | Cattle & buffalo | For accounting of milk production |
| Weekly | Cattle & buffalo | As a basis for concentrate feeding |
| Monthly | Cattle | Progeny testing & sire evaluation for estimation of individual cow's yield |
| Bimonthly | Buffalo | |
| Once in six weeks | Buffalo | |
| Three times in a lactation | Cattle & buffalo | To estimate herd performance |

Monthly milk recording is the most commonly used method in FPR. The following are the general rules for carrying out monthly milk recording.

Monthly recording should theoretically start on the 20th day after calving. The 20th day is the midpoint of a month after discarding the first five days as colostrum period. (30/2=15+5=20). Due to reasons such as information not reaching in time and the day too highly crowded for adding another recording, the first recording can be done within a period 20 +/−5 days. The first milk recording must not be performed before the 5th and after the 75th day of calving.

The interval between two recordings must be a minimum of 26 days and maximum of 33 days; for justified reasons, only one interval up to 75 days (but not longer) can be accepted in each lactation; if this longer interval reaches 100 days, the lactation can be calculated but the irregular recording must be mentioned.

If the animal is found to have dried-off, at the recording visit, the date of the dry-off of that animal is fixed at 14 days after the date of the last milk recording when still in milk or the actual date as informed by the farmer.

The milk yield of an animal for one day is obtained by recording the animal during all the times it is milked during a 24-hour period. It can be two times, three times or four times. The monthly recording to obtain the yield of one full day is the most commonly used method and is known as the reference method. According to this method, the acceptable interval between recordings is a maximum of 33 days (average of all recording intervals in the lactation). If other methods of recording are used, the results should be mathematically related to the reference method and a relative reliability factor quoted. The symbols used to show the time of milking are given in Box I.12.

**Box I.12** Symbols used to denote the times of milking in a 24-hour period

| Times of milking in a day | Symbol |
|---|---|
| Once a day | 1 x |
| Twice a day | 2 x |
| Thrice a day | 3 x |
| Four times a day | 4 x |
| Continuous milking (e.g. robotic) | R x |
| Both milked and suckled | S x |

## Milk recorder

The milk recorder is the person doing the recording based on the directions of the field performance recording organisation. Factors such as density of cattle population, number of cows per farm, type of milk recorder and the dwelling pattern considerably influence the investment and running cost of milk recording service.

Milk recordings are carried out by:

- Official milk recorders of the field performance recording organisation who undertake all the recordings. This includes recordings undertaken by approved on-farm systems that are supervised by an official representative of the recording organisation and that cannot be manipulated by the farmer or his nominee.
- Farmer or his nominee.

The conditions that favour the official milk recorder or the farmer are given in Box I.13

**Box I.13** Choice of type of milk recorder

|  | Official milk recorder | Farmer |
|---|---|---|
| Cattle population | Dense | Thin |
| Cows per farm | Low | High |
| Dwelling pattern | Clustered | Spread |
| Dairy consciousness | Low | High |
| Unemployment | High | Low |
| Literacy | Low | High |

In certain situations a combination of the two systems is also practised where recordings are undertaken by the farmer or his nominee and by an official representative of the field performance recording organisation. Whoever does the recording must be trained in all aspects of milk recording, such as milk recording per se, sample collection, identification of newborn calves, farmer contact and reporting.

## Farmers' role

The farmer interested in participating in milk recording must accept the regulations of the field performance recording organisation of his country, identify his animals with the required method, milk record all animals of the herd or allow the official milk recorder to record. Farmers when engaged for recording have to be provided with a booklet and an approved measure (balance or volume measure) for recording the milk on dates prescribed.

## Calculation of lactation yield

There are various methods to compute the total lactation production. The most common ones are described below.

### Test period method

The number of days for which a recorded day's production is represented is calculated and called the test period. For this the intervals between recordings are calculated. The first interval is the interval from the date of calving to the first date of recording. The second interval is between the first and second recordings, third interval between second and third recordings and so on. From these intervals the test periods are calculated. The first test period is the sum of the first interval and half days of the second interval. The second test period is the sum of the second and third intervals divided by two, third sum of the third and fourth intervals divided by two and so on. The 10th test period is the sum of half of the 10th interval and the number of days required to complete 305 days lactation or the date of drying, whichever is earlier. In the case of animals dry before 305 days, the last test period is the sum of half the last interval and the number of days from the last recording to the date of last milking (the day before the date of drying). Each test period is multiplied with the corresponding day's milk yield to get the test period production. The sum of the test period days is the number of days milked and the sum of the test period production is the quantity of milk obtained in that lactation. An example of calculation of the lactation yield from monthly records is given in Table I.6

**Table I.6** Calculation of lactation yield from monthly records—test period method

| Calving date 25.03.04 | | | Lactation ended on 03.01.05 | | | | | | |
|---|---|---|---|---|---|---|---|---|---|
| RD | ID | Milk kg | Fat% | Fat gr. | TP | TPM kg | TPF kg | Fat% |
| 8.4 | 14 | 29.4 | 3.72 | 1094 | 28 | 823.2 | 30.62 | |
| 6.5 | 28 | 24.2 | 3.51 | 849 | 29 | 701.8 | 24.63 | |
| 5.6 | 30 | 26.9 | 3.42 | 920 | 31 | 833.9 | 28.52 | |
| 7.7 | 32 | 24.0 | 3.61 | 866 | 29 | 696 | 25.13 | |
| 2.8 | 26 | 21.1 | 3.9 | 823 | 27 | 569.7 | 22.22 | |
| 30.8 | 28 | 18.3 | 4.12 | 754 | 27 | 494.1 | 20.36 | |
| 25.9 | 26 | 13.5 | 4.23 | 571 | 29 | 391.5 | 16.56 | |
| 27.10 | 32 | 9.8 | 4.58 | 449 | 29 | 284.2 | 13.02 | |
| 22.11 | 26 | 5.7 | 4.95 | 282 | 27 | 153.9 | 7.62 | |
| 20.12 | 28 | 4.8 | 5.16 | 248 | 28 | 134.4 | 6.94 | |
| Total | | | | | 284 | 5082.7 | 195.62 | 3.85 |

RD – Recording date; ID – Interval days; TP – Test period days; TPM kg – Test period milk kg; TPF kg – Test period fat kg

For cows milked three times a day the total of 3 milkings is considered as the day's yield and the fact '3 times milking' should be recorded under 'remarks'. Corrections should not be used for lactations that have naturally stopped. However, when animals are migrating from the recording area during lactation estimating standard lactation using correction factors is admissible.

When milk samples are tested for fat, the quantity of fat produced can be calculated by multiplying the fat percentage of the sample with the quantity of milk produced during the respective test period. The lactation average fat percentage is calculated by taking the total fat yield in kg as a percentage of the total milk yield in kg. For calculation of the protein yield and percentage the same method is used.

### Interpolation method or test yields method

In this method the intervals are multiplied with averages of the two milk yields. The following formulae are used to compute the lactation record for milk yield (MY), fat yield (FY), and fat per cent (FP):

$$MY = I_0 M_1 + I_1*\{.5*(M_1+M_2)\} + I_2*\{.5*(M_2+M_3)\} + I_{n-1}*\{.5*(M_{n-1}+M_n)\} + I_n*M_n$$
$$FY = I_0 F_1 + I_1*\{.5*(F_1+F_2)\} + I_2*\{.5*(F_2+F_3)\} + I_{n-1}*\{.5*(F_{n-1}+F_n)\} + I_n*F_n$$
$$FP = FY * 100/MY$$

where: $M_1, M_2, M_n$ are the weights in kilograms, given to one decimal place, of the milk yielded in the 24 hours of the recording day; $F_1, F_2, F_n$ are the fat yields estimated by multiplying the milk yield; and the fat per cent of the sample collected on the recording day; $I_1, I_2, I_n$ are the intervals, in days, between recording dates; $I_0$ the interval, in days, between the lactation period start date and the first recording date; $I_n$ the interval, in days, between the last recording date and the end of the lactation period.

The formulae applied for fat yield and percentage can be applied for other milk components such as protein and lactose. An example is given in Table I.7.

**Table I.7**  Calculation of lactation milk yield—interpolation method

| Calving date 25.03.04: | | Lactation ended on 03.01.05 | | | Sum | | |
|---|---|---|---|---|---|---|---|
| Rec. date | Interval days | Milk kg | Fat% | Fat g | Milk kg | Fat kg | Fat% |
| 8.4 | 14 | 29.4 | 3.72 | 1094 | 411.6 | 15.31 | |
| 6.5 | 28 | 24.2 | 3.51 | 849 | 750.4 | 27.20 | |
| 5.6 | 30 | 26.9 | 3.42 | 920 | 766.5 | 26.54 | |
| 7.7 | 32 | 24.0 | 3.61 | 866 | 814.4 | 28.58 | |
| 2.8 | 26 | 21.1 | 3.90 | 823 | 586.3 | 21.96 | |
| 30.8 | 28 | 18.3 | 4.12 | 754 | 551.6 | 22.08 | |
| 25.9 | 26 | 13.5 | 4.23 | 571 | 413.4 | 17.23 | |
| 27.10 | 32 | 9.8 | 4.58 | 449 | 372.8 | 16.32 | |
| 22.11 | 26 | 5.7 | 4.95 | 282 | 201.5 | 9.50 | |
| 20.12 | 28 | 4.8 | 5.16 | 248 | 147.0 | 7.42 | |
| | 14 | 4.8 | 5.16 | 248 | 67.2 | 3.47 | |
| Total | 284 | | | | 5082.7 | 195.61 | 3.85 |

### Centring method

This is an easy method of estimating the lactation yield, whose accuracy will go down with irregular and long intervals. In this method the sum of all the daily yields (separately for milk kg and fat gram) is multiplied by the average interval days. The average interval days are equal to the total days in milk divided by the number of recordings done. An example is given in Table I.8.

**Table I.8**  Calculation of lactation milk yield—centring method

| Daily yields kg | 29.4, 24.2, 26.9,24, 21.1, 18.3, 13.5, 9.8, 5.7,4.8 | Σ = 177.7 |
|---|---|---|
| Total days milked | (This is the day difference between the start of lactation and dry date, limited to 305 days) | 284 |
| Average interval days | 284/10 = 28.4 | |
| Total milk yield kg | 28.4 × 177.7 = | 5046.68 |
| Daily fat yield gram | 1904,849,920,866,823,754,571,449,282,248 Σ = 6856 | |
| Total fat yield kg | 28.4 × 6856/1000 = | 194.7 |
| Average fat % | 194.7 × 100/5046.68 = | 3.86 |

The total production figures calculated by the centring method differ slightly from those calculated by the other two methods.

### Estimation of lactation yield from part lactation

Methods to estimate lactation yields from part recordings are relevant where selection of animals has to be done before completion of lactation. One of the important uses of this method is to select bull calves from first calvers. Computation of lactation production from part lactation is also done when a large number of animals under recording migrate. However, short lactations due to disease conditions, normal drying, etc. should not be extrapolated to full lactation.

The standard lactation can be estimated form the peak yield and 100-day yield using the following thumb rules.

$$\text{Standard lactation} = \text{Peak daily yield} \times 200$$
$$\text{Or}$$
$$\text{100-day yield} \times 2.3$$

More appropriate factors could be developed for different breeds, seasons and geographical areas using a set of data from the corresponding population and analysed for these parameters. Factors are to be developed to compute the standard lactation from part lactations using the data available from the herds that are recorded daily using the following procedure.

- Take the records of around 100 animals that are recorded daily.
- Calculate average milk yields up to the 10th day, 20th day, etc. up to the 300th day after calving (30 averages). This can also be done at intervals of five days; in that case there will be 60 averages and 60 factors.
- Calculate the average standard lactation milk yield of the above animals from daily yields. This would be the sum of the daily milk yields for 305 days.
- Calculate correction factors by dividing the average lactation milk yield by the average milk yield up to the various stages of lactation.

Part lactation can be converted into a full lactation by using the appropriate factors. An example to work out correction factors is given in Table I.9. To apply correction factors take the milk yield of the animals to the nearest number of days and multiply with the corresponding correction factor.

**Table I.9**   Example to develop correction factors for correcting part lactation yields

The average milk yield of 100 cows in a herd for different periods of lactation is given in the column for yield up to days and the average standard lactation milk yield of these cows is 6128.8 kg. By dividing the lactation yield by the corresponding part yield the correction factor is obtained.

| Yield up to days | Milk kg | Factor | Yield up to days | Milk kg | Factor | Yield up to days | Milk kg | Factor |
|---|---|---|---|---|---|---|---|---|
| 10 | 179.7 | 34.106 | 110 | 2955.0 | 2.074 | 210 | 5414.5 | 1.132 |
| 20 | 419.0 | 14.627 | 120 | 3226.2 | 1.900 | 220 | 5544.9 | 1.105 |
| 30 | 669.8 | 9.150 | 130 | 3506.0 | 1.748 | 230 | 5662.2 | 1.082 |
| 40 | 944.3 | 6.490 | 140 | 3790.7 | 1.617 | 240 | 5767.3 | 1.063 |
| 50 | 1238.0 | 4.951 | 150 | 4078.1 | 1.503 | 250 | 5850.5 | 1.048 |
| 60 | 1529.0 | 4.008 | 160 | 4356.6 | 1.407 | 260 | 5918.2 | 1.036 |
| 70 | 1814.0 | 3.379 | 170 | 4622.5 | 1.326 | 270 | 5977.4 | 1.025 |
| 80 | 2097.0 | 2.923 | 180 | 4878.5 | 1.256 | 280 | 6025.3 | 1.017 |
| 90 | 2386.0 | 2.569 | 190 | 5096.5 | 1.203 | 290 | 6075.1 | 1.009 |
| 100 | 2672.2 | 2.294 | 200 | 5268.8 | 1.163 | 300 | 6114.4 | 1.002 |

The lactation milk yield of an animal that has produced 2345 kg in 120 days is 2345 × 1.9 = 4455.5 kg.

# Organisation of a field performance recording programme

In many developing countries a large proportion of the herds are very small and farmers do not find an immediate use/benefit in maintaining individual records, which makes the establishment of a field performance recording programme difficult. The objectives of the programme must be clearly understood by the implementers before launching the programme. Support and co-operation of the farmers can be secured through awareness creation and provision of some direct incentives at the outset. Financial support for cattle insurance, purchase of cattle feed and mineral supplements, health care and reproductive management are some of the incentives provided to farmers co-operating in field performance recording in India. Animal identification methods have to be agreeable to the owners, must be least painful to the animal and shall not cause bleeding.

## Monitoring of milk recorders

A schedule of animals to be milk recorded is to be prepared for each milk recorder indicating the dates on which each cow is to be recorded. While preparing the schedule, care should be taken to include optimum number of cows on each day of recording, to minimise the distance between cows recorded on the same day and to ensure easy supervision of milk recordings. Milk recorders engaged by the organisation are preferred in smallholder systems. Persons belonging to the village who can read and write are sufficient. He/she has to be given training to perform the assigned responsibilities. Experience in Kerala, India shows that milk recorders engaged on part-time contract are more efficient than regular employees of the organisation (Chacko, 1989). The part-time milk recorders must be properly informed of their duties and the terms of appointment. They have to be present at the time of milking of the cows and control all the cows as per the milk-recording schedule. When milk constituents are estimated, composite milk samples from morning and evening milk should be taken for analysis.

Regular supervision is the most effective among all the methods to ensure reliable data. A good supervisor will be a useful extension worker and an inspiration for the milk recorders who are doing a good job. General checking of recorded data and statistical evaluation are other means to check the correctness of the data. Large up-and-down fluctuations of daily yields of individual cows or exactly the same quantity of milk during most of the recording days can be the result of falsification. A large dataset should follow the rule of normal distribution. Where recording is done with an accuracy of 1/10 of a litre or kilogram, one can expect that the decimal figures from zero to nine should turn up with equal frequencies. The ratio between the observed and the expected can be determined using the chi-square test and the integrity of milk recorders with high chi-square values is often dubious. Chapter 2 of part II explains the methods of chi-square estimation. Persons who make mistakes and mischief should not be retained on the job.

## Financial aspects

Field performance recording requires funding for a long period, especially when it is done for progeny testing. Ensuring funds for the continued implementation of the programme is essential. The agency responsible for the genetic improvement programme of the area shall be responsible for the funding as well. It has been shown that the investment on the field performance recording programme in Kerala since 1978 yielded, on average, an annual return of 20% (Chacko, 1999).

# Information management

## Compilation and evaluation of data

Each milk recorded cow produces a large quantity of data, e.g. calving date, calf sex, owner, AI centre, monthly milk yields, fat percentages and so on. It is to be posted systematically and regularly for each animal in separate cards and lactation records have to be calculated. Necessary formats are to be designed for this at the beginning of the programme itself. Data maintenance and evaluation becomes efficient, fast and cheap when electronic devices are used. When the recording results are to be used for progeny testing of bulls, additional data to get information of breeding and environmental influences that may affect the breeding value (geographic area, year of calving, management, feeding, breed of dam, age at calving, sex of calf (if allowed to suckle), season of calving, lactation number, etc.) should be collected. The lactation yields are analysed using appropriate statistical techniques depending on the purpose for which the data is used.

## Feedback

A regular feedback of information about their cows' performance should be given to the farmers. This can be in the form of completed lactation reports of the cow under milk recording, a comparative statement showing the milk yield of all cows completing lactation during a particular period in an area or list of the top-ranking cows in the area and with the rank of the cow in question. The farmers must also be advised about a cost-effective system of feeding and management based on the data collected from their farms. All such feedback will benefit the farmer for scaling up his management practices and promote a healthy competition among the farmers.

# Conformation recording

Recording the conformation of animals in a herd is often done in the field as defined by the responsible international organisation, mainly for the purpose of breed conservation and development. Scorecards are made giving weight to each of the phenotypic characters that are true to the breed. The animal under recording is checked and scored. All traits must be scored or measured linearly from one biological extreme to the other. The range of scores must be from 1–9. Features that indicate the disposition for a genetically inadmissible defect should be highlighted. Dry cows cannot be considered for scoring especially when it concerns the udder and teat. Common norms for scoring should be evolved and persons responsible for scoring trained accordingly.

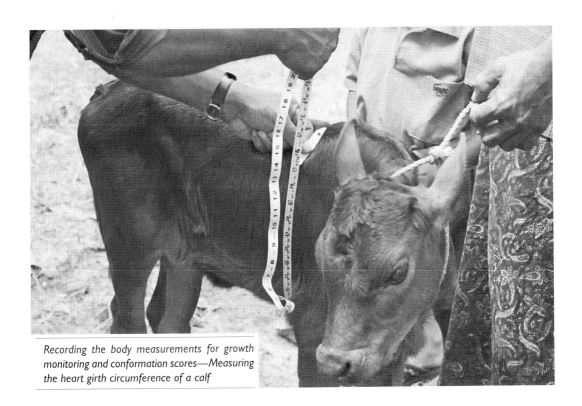

*Recording the body measurements for growth monitoring and conformation scores—Measuring the heart girth circumference of a calf*

# 7. Progeny testing of bulls

*A discussion of bull progeny testing was kept for the last chapter of Part I on purpose. Progeny testing is only successfully possible if quality bulls are available, good quality frozen semen is produced, a functioning artificial insemination service is in operation and a well-managed and carefully monitored field performance recording system is in place. This chapter discusses the various steps involved in progeny testing and quantifies them with examples. The genetic gain expected by progeny testing is described for various systems practised in the developed as well as in the developing world.*

# Introduction

Crossbreeding the local cattle that are generally non-descript with exotic dairy breeds is a breeding strategy for milk production enhancement in many tropical countries. The main objective of a planned crossbreeding programme is to combine the favourable *Bos taurus* and *Bos indicus* characters and exploit heterosis between them (Mason, 1974; Alstrom, 1977; Kunzi and Kropf, 1986). The breeding policy in such cases consists of two steps, building up a composite population with a given level of inheritance of the *Bos taurus* and the selection within the composite population. Among the many crossbreeding programmes in the tropics, systematic selection is practised only in few cases such as the AMZ breeding programme in Australia (Alexander and Tienery, 1990) and the Sunandini programme in Kerala (Chacko et al., 1985; Chacko et al., 1988; Jose and Chacko, 1992).

It is estimated that 77% of the genetic progress accruing in a population comes through the genetic paths cow to bull and bull to bull (Rendel and Robertson, 1950). Similar trends are also noticed in the zebu cattle of India (Singh and Nagarsankar, 2000; Gurnani, 1981). The breeding value (BV) for milk production of a bull, can be estimated most reliably from the mean milk production performance of a sample set of its progeny; the process is known as progeny testing (PT) and selection based on progeny testing as progeny selection. It can be shown that the average milk yield of five daughters recorded under uniform management will give an estimate of their mother's breeding value with equal reliability as estimated on the basis on its own production record (Lush, 1950). However, estimation of BV of an animal on the basis of one or few progeny may be misleading, due to the sampling nature of inheritance. The quantitative traits are influenced by environmental factors and all the progenies from the same bull will not receive the same environment. When large numbers of progeny are considered, these environmental effects get cancelled/balanced and a more reliable estimate of the breeding value is obtained. Progeny testing is the most commonly used method for estimating the BV of the bulls used in dairy production systems in developed countries. Farmer's herds are used for the purpose. Small herd sizes, low yield of the cows and non-availability of good systems are some reasons for non-sustainability of field progeny testing programmes in less developed countries.

There are two important sources of error in progeny testing: randomly distributed ones caused by segregation of inheritance or due to the sampling nature of inheritance and systematic errors (age effect of the cows, selection practised on the dams, etc). Increasing the number of progenies per bull can reduce these types of errors. The environmental differences between progeny groups raised in different herds during different periods of time also act as a source of error. This is reduced by adjustment of data for environmental factors.

# Steps involved

Decisions on number of bulls tested in a year, test herd, test AI, registration of the daughters, milk recording and data evaluation are the important steps involved.

## Bulls tested in a year

Every year a certain number of bulls have to join the stud as replacement for old bulls culled and disposed of and this number is related to the annual replacement requirement of breeding bulls and selection intensity. Their genetic quality depends on selection intensity. The number of bulls to be progeny tested is proportional to selection intensity. When selection intensity is less than 100, the number of bulls tested needs to be higher than that of the young bulls required for replacement. The number of young bulls to be tested every year with varying selection intensities to obtain 100 bulls for replacement annually is given in Table I.10. Table 1.10 shows that with increasing selection intensity the number of bulls to be progeny tested increases.

**Table I.10** Young bulls to be progeny tested annually to get 100 replacement bulls per year under varying intensities of selection

| Selection intensity (%) | 10 | 25 | 50 | 100 |
|---|---|---|---|---|
| Young bulls tested/$y^{-1}$ | 1000 | 400 | 200 | 100 |

In composite breeds formed from a crossbred population between *Bos indicus* and *Bos taurus* it was shown that young bull programme in which genetic progress reached the population through the bulls to breed bulls, proved to be an effective tool for genetic progress (Chacko, 1985). In this situation the number of bulls tested will be equal to the number of replacement bulls.

## Test herd

In large farms, as seen in dairying countries, some animals in each farm are allotted for test mating on mutually agreed terms between the farmer and the artificial insemination organisation. In small farm conditions, with only one or two cows in each farm, this system is not possible. Here all animals except bull mothers have to be inseminated with test bulls. The herd to be controlled under the progeny testing programme in small farms is related to the number of bulls to be tested, the coverage of artificial insemination, average progeny group size per bull and the fertility rate of the population. Herd size increases with increase in number of test bulls and increase in the progeny group size. With increased AI coverage and with a higher fertility rate the herd size will decrease. Exercise 5 given in the toolbox of the attached CD will help in calculating the herd size for progeny testing in small farm conditions with varying number of test bulls, progeny group size, fertility rate and AI coverage.

*Exercise 5.    Herd size for a field progeny testing programme*

As far as possible such a herd should be available in a contiguous area where the population density is high. The test herd should represent the target population so that the breeding values estimated are replicable in the target population. There should be good coverage of artificial insemination in the area and arrangements for follow-up of conception and calf birth.

## Progeny records

All test inseminations done have to be recorded, the conception rate checked and bulls with poor fertility excluded from the programme early. All the female calves born have to be identified so that optimum numbers of completed lactations are obtained.

### Registration of daughters

All efforts are to be made to include all the female progeny in the programme for which they should be identified early in life and followed up on a regular basis. It is important that female progeny of all the bulls are available. Farmer support programmes are used for retention of the female progeny in the test area (Trivedi, 2000). Care has to be taken to complete the programme as early as possible, which in turn results in a shorter generation interval and thus in a higher genetic gain per year.

### Milk recording of the progeny lactations

In organising a milk recording service under field conditions it is necessary to decide on the type of milk recorder. This may be the owner or a paid employee. Chapter 6 elaborated on which of these two categories is suitable in different situations. All progeny lactations should be enrolled for milk recording and there should be no type of selection, as this will affect the accuracy of the estimated breeding value of the bull.

## Monitoring

A close monitoring of milk recording should be done from the very beginning of the programme and unreliable milk recorders should not be allowed to continue. The number of persons required to carry out the milk recording (when paid employees are employed) will depend on the spread of the recording herd. Generally a person can work 25 days a month. Milking time being the same in all the farms in a village, attending to the milking of more than two farms in a day is not practical. Weighing the milk has to be accurate. However, getting a suitable and handy weighing balance is not all that easy. Each milk recorder needs to have a kit containing items such as weighing balance, sample dipper, sample bottles and reporting formats. Box I.14 shows the formats used in Kerala for the field progeny testing programme.

**Box I.14**  Milk recording formats used in the Kerala progeny testing programme

| 1. Milk recording schedule | | | | | | |
|---|---|---|---|---|---|---|
| Sl. No. | Date | Order of recording | Owner | Cow No. | Calving date | Last recording date |
|  |  |  |  |  |  |  |
|  |  |  |  |  |  |  |
|  |  |  |  |  |  |  |

Note: This is a schedule given to individual milk recorders and is their monthly job chart

2. Milk recording report

Name of unit                                             Name of milk recorder

| Record-ing date | Cow no. | Calving date | Owner | Day's yield | | | Sample No. | For lab use | | |
|---|---|---|---|---|---|---|---|---|---|---|
|  |  |  |  | M | N | E |  | Fat % | Protein % | Remarks |
|  |  |  |  |  |  |  |  |  |  |  |
|  |  |  |  |  |  |  |  |  |  |  |
|  |  |  |  |  |  |  |  |  |  |  |
|  |  |  |  |  |  |  |  |  |  |  |

Note: The milk recorder enters his daily milk recording work in this format, in duplicate. One copy is sent along with the sample to the lab and the other copy submitted weekly to the supervisor.

The details of individual cows are posted on separate sheets for manual operation or in individual cow files when personal computers are used. There are two possibilities:

- A file is opened in the computer for each cow at the time of starting the milk recording and information added to it.
- The data pertaining to a cow are manually recorded in a separate card and after completion of lactation is entered in the computer.

The estimated lactation yields can be sorted according to sire, dam, date of birth, date of calving, days milked, milk kg, and fat kg for making appropriate correction factors for all environmental effects. After identifying the factors that significantly affect the lactation yield, the daughter lactations are corrected using appropriate correction factors. There are many advanced methods for the estimation of breeding values wherein the corrections are carried out without bias to any of the factors. Least square analysis, best linear unbiased prediction and animal model are the popularly used techniques for breeding value estimation.

# Proven bulls

The whole exercise of progeny testing is to identify superior bulls and use them to produce the next generation. Once the breeding value is estimated 50% of the bulls will have a positive BV (they are above the average) and the other 50% will have a negative BV.

The genetic superiority of the sires can be used for improvement of the next generation using the paths bulls to breed bulls and bulls to breed cows. Chacko (1980) has shown that in Kerala, India, the young bull programme which employs the paths bulls to breed bulls and cows to breed bulls is as good as a normal progeny testing programme wherein the 50% best bulls of all tested bulls are used for artificial insemination. The top ranking (say top 10%) bulls are called 'proven bulls' and are generally used for nominated mating of bull mothers to produce the next generation bulls.

The genetic superiority of the bulls used for artificial insemination reaches the population through their daughters and sons. When young bulls are used for AI the generation interval is reduced. The earlier in the life of the parents their offspring are born, the shorter the generation interval. It is the goal of any breeding programme to strike a balance between the selection response and the generation interval so that optimum annual genetic gain is brought about.

# Time taken to get the sire proof

Zebu cattle and buffaloes are late to mature and their age at first calving as well as age at first semen collection are higher than the dairy breeds of the temperate zones. These aspects increase the time taken to get the sire proof and more so in low input dairying systems. Chacko (1989) reported that the average age of the bulls when the required doses of semen are ready as 39.2 months and the average age of the cow at first calving as 42.3 months.

The following elements constitute the time required to estimate the breeding value of a bull:

- Average age of bulls when the required doses of semen are produced
- Time required for completing the test inseminations
- Average age at first calving of the progeny
- Time taken to estimate the breeding value

Exercise 6 given in the toolbox included in the attached CD can be used to calculate the time taken to estimate the breeding value of the bulls under different situations.

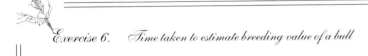

*Exercise 6.* *Time taken to estimate breeding value of a bull*

## Accuracy of the progeny test

Accuracy of the progeny test is a function of the heritability of the trait and the number of daughters screened. Table I.11 gives the accuracy of the progeny test changing with different numbers of daughters and heritability.

**Table I.11** Change in accuracy of the progeny test (as %) according to changes in heritability and number of daughters

| Heritability | Average number of daughter lactation per bull | | | | | | | |
|---|---|---|---|---|---|---|---|---|
| | 10 | 20 | 30 | 40 | 50 | 60 | 100 | 200 |
| 0.1 | 20.0 | 33.9 | 43.5 | 50.6 | 56.2 | 60.6 | 71.9 | 83.7 |
| 0.15 | 28.0 | 43.8 | 53.9 | 60.9 | 66.1 | 70.0 | 79.6 | 88.6 |
| 0.2 | 34.5 | 51.3 | 61.2 | 67.8 | 72.5 | 76.0 | 84.0 | 91.3 |
| 0.25 | 40.0 | 57.1 | 66.7 | 72.7 | 76.9 | 80.0 | 87.0 | 93.0 |

Figure I.17 shows that the increase in accuracy of the progeny test is not proportional to the increase in the number of daughters and the kind of increase obtained with more than 50 to 60 completed lactations is not commensurate with the cost involved. The herd size is related to the number of daughters to be averaged per bull.

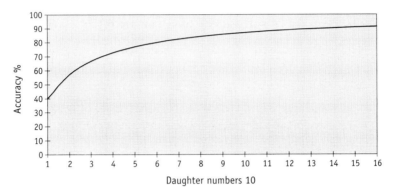

**Figure I.17**   Change in accuracy of PT according to changing daughter number

**Test insemination**

Artificial insemination units in the selected area carry out the required number of test inseminations. The number of test inseminations required to have a certain number of completed first lactation of the daughters will be related to the average number of inseminations required per calf born; sex ratio and the percentage of first daughter lactations obtained from the female calves born. It was shown that an average of 29.4 inseminations are required to produce a completed first lactation in Kerala (Chacko, 1992). He suggested that intensive control of the test herd, awareness creation and farmer support can reduce the ratio between test inseminations and the completed first lactations.

Ideally daughter lactations of all test bulls should appear in all areas and through all the seasons in equal numbers so that environmental effects are balanced. This must be the objective while planning the test inseminations. One possibility is to distribute the semen of each bull through all the artificial insemination units and seasons, divide the number of semen doses of each bull by the number of seasons and distribute one portion of semen of all the bulls through all the AI units, then take the second portion and so on until all the semen doses are exhausted.

## Genetic gains through proven bulls

There are two possibilities of utilising the genetic superiority of the bulls to improve the population.

### Orthodox progeny testing programme

Based on the breeding value estimated from progeny performance select a certain percentage of the bulls and use them to produce the next generation males and females. This means that only selected bulls are employed in large-scale artificial insemination. Here:

- We produce semen from more bulls than what is required for AI (depending upon the selection intensity)
- Use it on a limited number of cows to produce the required progeny lactations for sire evaluation
- Store all the semen produced by the bulls until the breeding value of these bulls is available
- Select the required proportion of top ranking bulls
- Use their semen for large-scale AI
- Discard semen doses from the rest of the bulls

### Young bull programme

In this programme a small percentage of bulls are selected according to the breeding value and used to produce the next generation bulls. Here simultaneous with the test AI programme the young bulls are used for large-scale artificial insemination. The superiority of these proven bulls is brought into the population only through their sons.

It is shown that the added genetic improvement achieved through the orthodox progeny testing programme will not be much when the generation interval is long as is the case with crossbred populations (Chacko, 1985). At the same time, the infrastructure and financial needs of the former is three- to fourfold of the modified progeny testing programme. The implementation of the programme will be basically similar in both cases. We know that a certain number of bulls is to be replaced every year. In a young bull programme where no selection is applied for bulls used for large-scale artificial insemination, the replacement bulls can be put to progeny testing and a certain proportion of them selected as bulls to produce the next generation bulls.

Progeny testing can be done either in a large organised farm or in the field. The on-farm testing requires a large farm and the older cows in the farm have to be periodically removed to make space for the young cows. More than the requirement of a large farm and its running, this has a major genetic disadvantage. When there is a genotype by environmental interaction, the results of on-farm testing will not 'hold true' in the field, since the environment is different. In most of the dairying countries progeny testing is done with the herds available with the farmers. The main challenge in such programmes remains the establishment and continuance of a reliable field performance recording system.

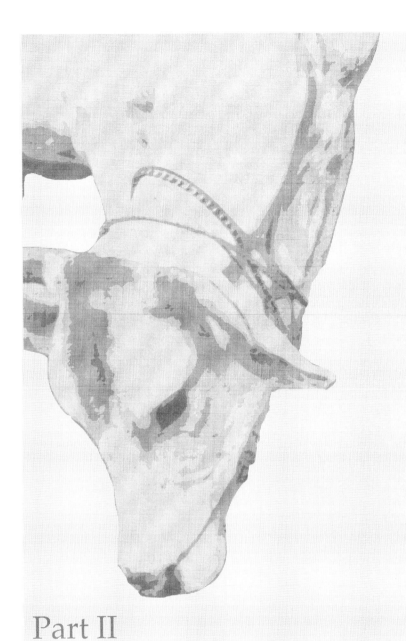

# Part II

# Genetics and Statistics for Practical Breeding Work: A 'Hands-On' Approach

*In spite of the abundance of standard literature on genetics and statistics relating to livestock sciences, this book without a discussion of basic principles in genetics and statistics, would not be comprehensive. Part II commences with genetic principles and continues with the statistical basics necessary to analyse breeding data for selection and breeding decisions leading to a genetic gain. These principles are illustrated by case studies and with corresponding examples in the toolbox, which allow the reader to practise these principles with her or his own set of data.*

# A. Quantitative Genetics

*The principles of quantitative genetics are summarised based on various practical examples. The sources of variation are discussed and are adequately illustrated. These principles are taken further to the discussion and illustration of selection methods and the estimation of breeding values. The breeding value estimations are supported with two toolbox exercises. The chapter continues to describe the breeding systems such as inbreeding, crossbreeding and upgrading, making use of the genetic principles discussed earlier.*

# Introduction

Not surprisingly we observe that calves born to Holstein Friesian parents are black and white in colour. Whatever the changes made in the management of these calves the coat colour will not change. Genes control many characters like the coat colour of a living being. Genes are maintained in the population and are passed on from one generation to the next. They are located at particular points called loci (singular form, locus) in the chromosomes. The number of chromosomes is fixed for the species. Chromosomes as well as genes occur in pairs.

Genes can have different effects on the character and behave in three possible ways in relation to each other; dominant, recessive or additive. When the action of one allele dominates over the other the dominating one is called dominant and the one dominated, the recessive allele. Conventionally capital letters are used to denote dominant alleles and small letters for recessive alleles. It is also possible that the two alleles are equal in power so that the phenotype of the heterozygote will be half-way between those of the two homozygotes. This type of gene action is called additive gene action.

The alleles of any one pair may behave differently depending on the influence of each on the character:
The phenotypic expression of a character is the combined effect of genes, environment and possibly an interaction between the two. This means that the influence of genes cannot be directly identified from phenotypic expression of the trait.

A large amount of data on a specific quantitative trait plotted as a frequency distribution in the form of a histogram will take the shape of a bell. (Figure II.1). The distribution of a large dataset of a quantitative trait follows a normal distribution.

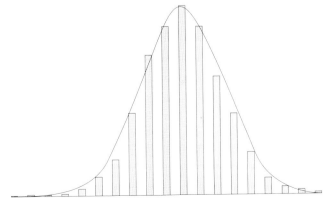

**Figure II.1**    Normal distribution curve

# Sources of variation

## Qualitative and quantitative characters

The coat colour referred to earlier is a *qualitative character* and normally one pair of genes controls such characters. Expression of qualitative character is in discrete classes and they often differ in kind. Examples of qualitative traits are horned or polled, skin colour, albino or pigmented.

Many pairs of genes control most production traits and the expression is on a continuous scale. Such characters are referred to as *quantitative characters*. Examples are milk production, fat percentage in milk, growth, etc. Quantitative characters:

- Show continuous variation
- Many pairs of genes control them
- Environment plays a significant role in the expression of the trait
- Actions of genes are usually additive.

There are two possibilities for the breeder to improve genetic properties in a population

- Choice of individuals to be used as parents of the next generation—Selection
- Control of the way parents are mated—Mating systems

The process of breeding and selection in large ruminants may be envisaged in four paths for transmission of genes from one generation to the next (Figure II.2).

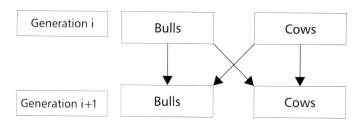

**Figure II.2** The four paths of gene flow

The present generation is denoted as generation i, and the subsequent generation as generation i+1. Bulls used in artificial insemination or natural service in generation i will be sires of the next generation bulls and cows. So also cows of $i^{th}$ generation will be dams of the $i^{th}$ +1 generation of cows and bulls. This is a continuous process and often generations overlap. In dairy cattle when AI is employed as a means of breeding, the genetic progress achieved through the aforesaid paths is in the proportion (percentage) given below (Rendel and Robertson, 1950).

- Bulls to bred bulls        43%
- Bulls to bred cows        18%
- Cows to bred bulls        33%
- Cows to bred cows          6%

## Genetic and environmental sources of variation

Differences between individuals with respect to expression of the trait give rise to variation, which is the key to making genetic change. Selection is possible only when there is variation among individuals. If all animals have the same measure of a trait, it would not be possible to select for the trait.

An animal's phenotype is determined by its genetic and environmental influences:

$$P = G+E$$

where
P—Phenotype, the way the animal presents itself
G—Genotype
E—Environmental influences

Phenotypic measurements of a quantitative trait vary between individuals (phenotypic variance) due to differences among them with regard to environment (feeding, management, health status, climatic conditions, etc.) and differences in their genetic make-up. Phenotypic expression of a trait is the one measured in individuals. Phenotypic variation can be easily calculated from recorded observations. It consists of parts due to heredity and due to environment. To be more precise total phenotypic variation is a function of genetic variation and environmental variation.

$$V_P = V_G + V_E$$

where:   $V_P$ is Total variation (phenotypic);
         $V_G$ is Genetic variation (due to gene action);
         $V_E$ is Environmental variation.

Genetic variation can further be split into:
- Variation due to additive gene action—*additive genetic variation*—$V_A$. Additive genetic variation is the most important kind of genetic variation.
- Variation resulting from dominance gene action—*dominance genetic variation* – $V_D$.
- Variation due to epistatic effects – *epistatic variation*—$V_I$

$$V_G = V_A + V_D + V_I$$

Taking into account the above partitioning of the genetic variation, it is now possible to explain phenotypic variation as:

$$V_P = V_A + V_D + V_I + V_E$$

The amount of variation is measured and expressed as variance. In animal breeding practice it is often difficult to measure with precision the variances due to dominance and epitasis. Therefore, for all practical purposes we are effectively left with additive and environmental variances. The non-additive genetic variances are included in the environmental variance. So for calculations in applied animal breeding, the formula will be:

$$V_P = V_A + V_E$$

Phenotypic variance can be partitioned into additive and environmental variance by comparing groups of animals that are related (they have some genes in common). Comparing differences between twin pairs with those of non-twin pairs will provide an estimate of genetic variation. If twins were very much similar in a particular trait and the other individuals were highly different in the same trait, this would indicate that heredity has a very high influence on the trait. But twins are not often available in large numbers in cattle and buffaloes, and as such do not offer a solution to estimate variation due to genes and environment. In genetic studies parent offspring, full-sib and half-sib analyses are most commonly employed for estimation of genetic variance. This is further explained under heritability below.

## Heritability

The relative importance of heredity in estimating phenotypic values is called heritability. Mainly it is the additive genetic effects that are passed on from parents to progeny; hence it is important to assess it. This can be done indirectly from an assessment of the phenotypic variance. The ratio of the additive genetic variation to the total phenotypic variation is called heritability and thus is the term used to describe the strength with which a quantitative trait is inherited.

$$\text{Heritability} = V_A / V_P \text{ and is denoted by } h^2$$

Heritability is generally expressed as a proportion or percentage. This would mean that heritability is the proportion of total variation among a group of individuals measured due to additive genetic variation. This definition implies that heritability is a measurement and not a characteristic. For example we do not estimate heritability of 'milk yield' but rather the standard lactation milk yield of first calvers in India during 1999 (a specified set of data). This is the reason why there are several estimates of heritability for the same trait. However, there is no reason why different populations should contain more or less the same amount of genetic variance. This is especially true with populations having very long breeding histories (e.g. established breeds of cattle). But in the case of recently developed breeds (e.g. synthetic breeds, crossbreeds) such heritability estimates may not be the same and are often smaller due to very high genetic variation. Heritability is population dependent. Usually, $h^2$ is estimated from a sample or subset of the population and inferences drawn about the population.

The ratio of total genetic variance ($V_G$) to total variance is called heritability in the broad sense whereas $V_A$ / $V_P$ is called heritability in the narrow sense. In this book wherever heritability is referred to, heritability in the narrow sense is meant.

The higher the heritability, the better the response to selection. Heritability estimates around 0.1 (10%) or below are regarded as low, those between 0.1 and 0.3 (10 – 30%) medium and above 0.3 (30%) high. Traits associated with reproduction and survival have low heritability, milk production and early growth rate traits medium heritability and various qualitative traits high heritability. Heritability estimates of some important traits in dairy cattle and their correlation are given in Table II.1.

**Table II.1**  Heritability estimates of some traits in dairy cattle

| Trait | $h^2$ | Correlation between traits |
|---|---|---|
| Whole milk production | ** | Milk content |
| Milk content (fat%, protein%) | *** | Milk yield |
| High roughage intake and ability to digest | *** | Not yet known |
| Longevity | * | ? |
| Fertility | * | ? |
| Health | * | ? |
| Low age at first calving | *** | Lifetime milk yield |

* low  ** medium  *** high

## Estimating heritability

Partitioning of the total phenotypic variance into additive, genetic and epistatic variation can be done by comparing groups of animals which share some genes in common but not all. Differences between the performance of two breeds maintained as a part of a single herd under the same feeding and management conditions will represent a real genetic difference between the breeds in question. Providing uniform management will reduce the difference between performance of animals in a group and thereby the phenotypic variance. Phenotypic variation being the denominator in the ratio to estimate heritability will in turn increase the heritability estimate.

Heritability of a trait can be estimated rather easily using:
- Identical twins
- Correlation between dam and daughter
- Half-sibs

More complex statistical methods are available to estimate population parameters such as heritability, but these methods are not covered here.

## Identical twins

Identical twins have all their genes in common where as heterozygotic twins on average share half of the genes they received from their parents. There may be differences in performance regarding a trait in question within twin pairs. Also there may be differences between the various twin pairs examined. Through variance analysis it is possible to partition these differences as given below. Variance between ($V_B$) twin pairs is the sum of environmental variance and twice the additive genetic variance whereas variance within ($V_W$) twin pairs is the environmental variance.

|  | Degrees of freedom (df) | Expectation of mean square deviation |
|---|---|---|
| Between twin pairs | $n - 1$ | $V_W + 2V_B$ |
| Within pairs | $n$ | $V_W$ |
| Heritability | $V_B / V_W + V_B$ | |

Twins are not available in large numbers in dairy cattle. They also have the disadvantage of sharing a common environment in the uterus and during the early period of their life. As such they are not preferred in dairy cattle for heritability estimation.

### Regression of offspring on parents

Within sire performance of the daughter can be compared with that of the dam. However, this should be done at the same stage, e.g. if lactation yield is to be compared, both mother and the daughter yields should be of the same lactation. Further, refinement of the data by correcting for effects of year and season would also be necessary. The expected regression of daughter yield is $\frac{1}{2}h^2$. Therefore $h^2 = 2 \cdot b$ (with b as regression coefficient of daughter yield on dam's yield.)

### Correlation of half-sibs

Paternal half-sib analysis is the most common method for estimating heritability of milk yield in dairy cattle. It is often readily available when field performance recording programmes are taken up for the purpose of sire evaluation. Half-sibs, on average, share 25% of the genes. For the estimation to be valuable two conditions should be satisfied: that the sires are a random sample of the breed and that all individuals, within the analysis, have the same management opportunities.

In a situation wherein the progeny group size is equal, variance analysis to estimate heritability would be:

|  | df | Expectations of mean square |
|---|---|---|
| Between sires | $S - 1$ | $V_W + nV_S$ |
| Within sires | $S(n - 1)$ | $V_W$ |

Here: S is the number of sires and $n$ = number of daughters per sire; $V_S$ the Variance between sires.

The variance component between sires ($V_S$) is equal to $\frac{1}{4}V_A$ and the variance component within progeny groups ($V_W$) is equal to $V_P - \frac{1}{4}V_A$.

$$h^2 = 4V_S / (V_W + V_S)$$

Equal progeny group size is not commonly available and in order to account for the unequal progeny group size, $n$ has to be replaced with $[\Sigma n - \Sigma n^2 / \Sigma n] / (S - 1)$

## Genotype – environment interaction

There are situations in which heredity and environment interact. Examples are seen in dairy cattle breeding. Breeds compared in one environment may express certain productivity. But the same breeds compared in another set of conditions would behave differently. Holstein Friesians may perform better than Zebu breeds in a tropical environment, but are nowhere near to their production levels in temperate conditions. Zebus will probably perform better in a better environment, too. Thus, genotype $x$ environment interaction exists only when the ranking is reversed or when the differences between good and bad environment are different between the two groups. Standard lactation milk yield of Friesian cows whose sires originated from 10 countries, were found to very highly in two different management levels (Table II.2).

**Table II.2**  Milk yield (kg) of cows originated from sires in different countries under two management levels

| | USA | Israel | Canada | New Zealand | Sweden | UK | Denmark | Germany | Poland | The netherlands |
|---|---|---|---|---|---|---|---|---|---|---|
| Field trial | 3783 | 3713 | 3695 | 3576 | 3438 | 3378 | 3371 | 3332 | 3265 | 3256 |
| Intensive trial | 5402 | 5222 | 5225 | 4996 | 4909 | 5065 | 4706 | 4933 | 4397 | 4539 |
| Difference% | 30.0 | 28.9 | 29.3 | 28.4 | 30.0 | 33.3 | 28.4 | 32.5 | 25.7 | 28.3 |

*(adapted from Reklewski et. al.1984)*

The increase in milk yield obtained by putting the animals on intensive management ranged from 25.7 to 33.3%. This example shows expression of genotype – environment interaction by different groups of animals. Exotic breeds are often imported to tropical low input systems and invariably the milk yield is far below expectations. In the tropics nucleus herds (usually maintained in organised farms) are normally maintained in a better environment than target herds (often maintained as small herds by farmers), on average kept in lower input systems of management.

The discussion of genotype – environment interaction can well be a discussion about which breed will do best under specific circumstances. Adaptation to local environments is then a particular example of genotype – environment interaction.

# Selection

## Principles of selection

Selection in the context of livestock improvement is defined as a process of giving preference to certain individuals as parents of the next generation. In genetic terms, it is a process of differential reproduction of genotypes. The process of selection operating through choice of individuals depends on two forces.

**Natural selection:** Its principle is 'survival of the fittest' in a given environment. Individuals who are 'not adapted' or 'not fit' are either eliminated due to death or fail to reproduce, resulting in extinction.

**Artificial selection:** This force of selection is man made and depends upon the will or choice of the breeder, who decides which animals are to be retained as parents to produce the next generation. The choice of the breeder is objective specific.

These two selection forces may act simultaneously on the same trait or on different traits and their action may be antagonistic whereby individuals preferred by the breeder may be less fit with respect to certain fitness traits. This undesirable genetic correlation between fitness and choice of the breeder results in less than expected response to artificial selection.

## Consequences of selection

The result of selecting superior animals as parent stock is a change in the mean performance of the progeny generation. Gene frequency of alleles carried in that population is thus changed. Frequency of desired genes is increased at the expense of undesired ones. In the normal course selection is exercised to improve the economic traits in domestic livestock and hence works on additive effects of genes. However, selection can be applied on simple qualitative traits as well.

Identification of superior individuals is a complex process and is completed in different stages. Genetic selection is based on breeding value or genetic worth of an individual. The breeding value is the sum of average effects of genes possessed by the individual. It can be estimated based on two criteria: individual performance and performance of relatives. Sources of information based on which the breeding value of an individual is estimated, are called 'basis of selection' or 'selection criteria'. The resultant BV obtained is known as estimated breeding value (or) probable breeding value. The basis for estimating the BV of an individual from the phenotypic values of its relatives lies in the fact that the individual and its relatives have some genes in common.

When the breeding value of an individual is estimated based on:
* Individual's phenotypic value, this is called individual selection.
* Performance of ancestors is known as pedigree selection
* Collateral relatives termed family selection
* Progeny, known as progeny selection (or) progeny testing

## Individual selection

The animal itself is the smallest unit of selection and improvement. Its own phenotypic value of the trait under consideration is used to estimate the breeding value of that individual. This method is:
* Comparatively easier

- Can be used when other information is not available
- Can be applied earlier to other selection methods

However it cannot be applied:

- In the case of sex limited traits
- When the traits are expressed in later life of the individual (or) after death of the individual
- For traits with low heritability.

In dairy cattle, particularly for male selection, individual selection is not very useful. However individual selection is of great value for selecting cows.

Individual selection when applied for more than one trait is known as multitrait selection. It is useful and important to estimate the breeding value of an individual considering several traits simultaneously instead of a single one, because the economic value of an animal depends upon several traits called the net merit of an animal. For example, a dairy cow which yields medium milk with medium fat and protein content for a longer duration and remains dry for a shorter period between two successive calvings is more economical than a cow yielding more milk with higher fat and protein content for a shorter duration, then going dry earlier and remaining dry for a longer period. There are three important methods of multitrait selection.

## Tandem selection

In this method, individual traits are improved successively i.e., selection is practised for only one trait at a time till satisfactory improvement in it is achieved. Selection is then done for the second trait and so on. This method is less efficient because the genetic progress per unit of time is less. Tandem selection is widely used by farmers however, because so simple.

## Independent culling levels

This method involves selection for two or more traits at a time. A minimum standard (level) is fixed for each trait under consideration. To arrive at the minimum levels for culling for each trait, several points are taken into considerations:

- Relative economic importance of the trait
- Heritability
- Correlation between the traits considered
- Biological requirement to maintain the population structure

Effectiveness depends on the cut-off levels or standards kept for each trait. If very low levels are kept many animals get selected and efficiency decreases. On the other hand, if high standards are set not enough reproducing animals are selected, which might result in a gradual extinction of the population. But the independent culling levels method is superior to tandem selection because selection is practised for more than one trait at a time and allows early culling of inferior animals. The disadvantages of this method are: it does not permit superiority in some trait to compensate for deficiencies in others and the procedure of calculating optimum culling levels is tedious.

One practicable use of this method is when the traits to be combined appear at different stages of life and correlate positively. An example in dairy cattle would be selection of female animals for growth and first lactation milk yield. There could be a cut-off point for culling female animals for substandard growth. Animals growing below the fixed level shall be eliminated and not allowed to come to milk production. Cows already selected for growth shall be subjected to selection again for milk yield based on the standards fixed.

### Selection index

In this type of selection, a score weights the individual traits and these scores are summed up for a number of traits to get the index value or total score. It is an index of the net merit of an animal for many traits. Here individuals are judged on the basis of a linear function of their measurements. Construction of the selection index is rather simple when the traits considered are not correlated. Generally the selection index constructed in one set of environmental conditions and management practices will not hold good in another set of conditions. Finding genetic and phenotypic parameters is not an easy task, especially when many traits are included in the index. The relative economic importance of traits varies from time to time and the relative economic value of each trait is often difficult to establish.

The relative gains per generation by employing the three different types of multitrait selection with varying numbers of characters and when the best 1% and 2% are selected assuming same heritability for a character in all types of selection is shown in Table II.3

**Table II.3** Relative gain per generation with three types of selection

| Proportion selected | Selection method | Number of characters | | |
|---|---|---|---|---|
| | | 2 | 4 | 8 |
| 1% | Tandem | 71 | 50 | 35 |
| | Independent culling | 93 | 84 | 74 |
| | Index | 100 | 100 | 100 |
| 2% | Tandem | 71 | 50 | 35 |
| | Independent culling | 87 | 77 | 65 |
| | Index | 100 | 100 | 100 |

(Source: Robertson, 1980)

## Pedigree selection

Pedigree means the ancestors of the individual, which may be parents (sires, dams), grand parents etc.. The pedigree information, which contains the phenotypic values of the ancestors with respect to economic traits, is important as far as the breeder is concerned. Pedigree selection is more accurate when the heritability of the trait considered is high. More emphasis has to be given to information on the immediate ancestors for more efficient pedigree selection. The practical difficulties in using pedigree selection are:

- Non-availability of pedigree information
- Faulty pedigree information
- Low heritability of many economic traits

This method is less costly, easier and is applicable in the case of traits not expressed in the individual or expressed at a later stage of its life. It is useful as additional information to individual selection and is helpful in multistage selection programme, wherein initial selection is based on ancestor information. The disadvantage is that all animals of similar pedigree are sometimes culled in spite of the fact that some of them are really good. Additionally, pedigree information can predict an animal's breeding value only with limited accuracy (maximum 50%), as differences between full sibs cannot be detected.

## Family selection

Full sibs and half-sibs are most commonly considered in this type of selection. The greatest advantage of family selection is that it can improve traits with low heritability values in species with higher reproduction rates. It is comparatively costly and leads to a higher chance of inbreeding. The accuracy of selection increases with increasing number of sib information and can be calculated for a halb-sib structure using the formula:

$$r = \frac{1}{2} \cdot \sqrt{n/(n + k)} \qquad \text{where } k = (4 - h^2)/h^2$$

Example: Number of half-sibs, $n$, = 50, $h^2$=0.25

$$r = \frac{1}{2} \sqrt{50/(50 + 15)} = 0.438$$

## Progeny selection

Estimation of breeding value of an individual based on the performance of its progeny is called progeny selection or progeny testing (PT). Progeny testing is regarded as a modified form of family selection and is considered as very accurate, especially in dairy cattle. It is useful in selection for traits having low heritability values and for sex limited traits. It is commonly said that the pedigree gives an indication what an animal should be, the individuality reveals what an individual looks like and the progeny reveals what an individual is. Progeny testing is also useful to prove whether a sire is free from any recessive genes of lethal/sublethal nature. The important limitations of progeny selection are the time, cost and long generation interval. Progeny selection in dairy cattle takes a long time because of the longer generation interval. It is possible that the individual may either be dead or become too old to reproduce by the time the results are available.

The accuracy of the progeny selection can be calculated using the formula:

$$r = \sqrt{n/(n + k)} \qquad \text{where } k = (4 - h^2) / h^2$$

Example: Number of progeny = 50, $h^2 = 0.25$

$$r = \sqrt{50/(50 + 15)} = 0.88 \text{ or } 88\%$$

## Factors affecting selection progress

Genetic progress as a result of selection is referred to as selection response and is defined as the difference between 'mean' of the progeny generation and 'mean' of the parental generation with respect to the trait(s) under consideration. The selection response is abbreviated as:

$$RS = (\overline{X}o - \overline{X}p)$$

where: RS is the response to selection;

$\overline{X}o$ the mean phenotypic value in the offspring generation;

$\overline{X}p$ the mean phenotypic value in the entire parental generation, before selection.

Response to selection can be measured only after the selection process has acted upon the population. Three important factors affect genetic progress through selection.

### Selection differential

The difference in the character selected between the selected parents and the group from which they are chosen is referred to as selection differential (Box II.1). It is the mean phenotypic value of the individuals selected for breeding, expressed as deviation from population mean. If there are variations between the parents with regard to the number of progeny they contribute to the next generation, then the mean difference, weighted according to number of progeny contributed, is called the realised selection differential.

It is normal to have different selection differentials for the two sexes. In cattle where AI is used, the selection differential for males would be far higher than for females. When considering the individual herd, flock or the system as a whole, the selection differentials from the two sexes have to be averaged (Box II.1)

**Box II.1**   Averaging selection differentials for first standard lactation milk production in cattle

| 1 | Choosing female replacements from dams | (dam – daughter) |
|---|---|---|
|   | Average of selected dams (best 50%) | 1720 kg |
|   | Average of herd | 1400 kg |
|   | Selection differential (for females) | 320kg |
| 2 | Choosing male replacements from dams | (dam – son) |
|   | Average of selected dams (best 5%) | 2224 kg |
|   | Average of herd | 1400 kg |
|   | Selection differential (for males) | 824 kg |
| 3 | Average selection differential | = (320 + 824) / 2 = 572 kg |

It is necessary also to consider the selection differential on the sire side, if known. This involves, corresponding to the above example, the difference from the average of sires chosen as fathers of females and sires chosen as fathers of male replacements. In this way there are 4 separate selection differentials to make up the average instead of the 2 used in Box II.1.

## Heritability

Heritability, being the ratio between additive genetic variance and phenotypic variance of the trait for which selection was exercised, affects the response to selection. As heritability is for a set of measurements and not for the particular trait, it is ideal to use heritability figures estimated from the population considered. It may not always be possible to do so in which case published estimates from a similar kind of population kept under similar environmental conditions would be the next best.

## Response to selection

Response to selection is a function of the selection differential, heritability of the trait and the generation interval. The selection differential can be predicted from two other pieces of information:

- The proportion of females and males which it is necessary to retain as parents of the next generation
- The amount of variation present for the trait under consideration

## Proportion selected

Selection differential is standardised by dividing it by standard deviation of the phenotype. This called as the standardised selection differential or selection intensity.

$$sd/\sigma_P = i$$
$$sd = i \cdot \sigma_P$$

Where sd is the selection differential and i the selection intensity.

It is a statistical convenience that for relatively normally distributed traits the intensity of selection (i) can be derived directly from the proportion selected. The corresponding figures can be looked up in tables prepared for the purpose (table within Figure II.3, Falconer, 1989) or, less accurately, from the graph in Figure II.3.

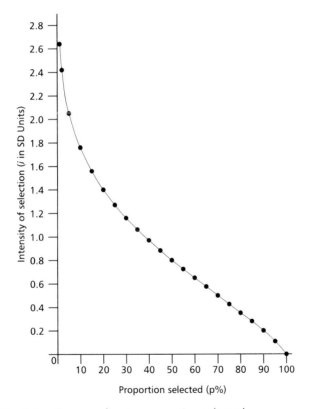

**Figure II.3**   Selection intensity according to proportion selected

| Proportions of animals selected (p%) | 100% | 90% | 80% | 70% | 60% | 50% | 40% | 30% | 20% | 10% | 5% | 2% | 1% |
|---|---|---|---|---|---|---|---|---|---|---|---|---|---|
| Value of i | 0.20 | 0.20 | 0.35 | 0.50 | 0.65 | 0.80 | 0.97 | 1.16 | 1.40 | 2.05 | 2.05 | 2.42 | 2.64 |

The figures for selection intensity are given in units of standard deviations. Thus, if the best 1% of animals are selected, they will be, on average, 2.64 standard deviation (SD) above the mean of all the animals in the distribution. Some other examples are: the best 5% = 2.05 SD; top 30% = 1.16 SD; and top 50% = 0.80 SD. If, in practice, it were possible to cull only the poorest 10% of the breeding females, the 90% retained would be a little better than the average—by approximately 0.2 SD units.

The genetic change that occurs as a result of selection (selection response; R) in the next generation is the product of the selection differential (intensity of selection · phenotypic standard deviation) and the heritability.

$$\mathbf{R} = \mathbf{i} \cdot \sigma_p \cdot \mathbf{h}^2$$

where: $h^2$ is heritability;
i the intensity of selection;
$\sigma_p$ the Phenotypic standard deviation of the trait

## Generation interval

Generation interval is defined as the average age of the parents when their offspring are born or, more precisely, those offspring used to replace the parents. The result of selection practised in one generation can be seen only in the subsequent generations. The genetic changes that accrue in one generation divided by the average generation interval is called average annual genetic gain. The faster the turning of generations, the faster the average improvement per year, other things being equal.

The generation interval is determined by:
- Age when the animals first start to breed
- Interval between successive parturitions
- Average number of years bulls are used for breeding
- Number of offspring born on each occasion that survives to breeding age.

The earlier in the life of the parent its offspring are born, the closer the parturitions follow each other and the more offspring per parturition, the sooner the number needed as replacements is reached, and hence the lower the generation interval.

Generation interval is found to be generally longer in the case of *Bos indicus* breeds of cattle and buffaloes. Poor feeding and environmental stress will adversely affect all of these attributes. Hence it is difficult to generalise the length of the generation interval.

Since generation interval affects the rate of genetic progress from selection, it is an advantage to shorten it as much as possible consistent with other requirements. Cautious attempts are necessary from the part of the breeders to reduce the generation interval to the shortest extent possible. Improving the reproductive management of the animals and use of breeding bulls for shorter periods are possibilities to bring down the average generation interval.

Generation interval is usually greater for female parents because they are kept to produce offspring for several years than is necessary for males of the species. However, though males do not usually have to be used for several years, they are often retained until their progeny test result is obtained (Box II.2).

**Box II.2**  Calculation of average generation intervals in cattle

The average age at first calving in a herd is 4 years, calving interval 1½ years and cows maintained for 5 lactations. In this situation the average generation interval of cows to produce the next generation cows (daughters) would be (4+5.5+7+8.5+10)=35/5 = 7 years (It is assumed that all the female calves born to these cows are needed as replacements.) Bull calves are kept only from the first calvers and hence the generation interval of cows to produce next generation bulls (sons) would be 4 years. The breeding bulls are used for AI at the age of 2.5 years and continuously for four years (2.5+3.5+4.5+5.5)/4= 4. Proven sires produce young bulls for replacement and the bulls are proven at the age of 9 years.

The four separate generation intervals are:

| | |
|---|---|
| Dam to daughter | = 7 years |
| Dam to son | = 4 years |
| Sire to daughter | = 4 years |
| Sire to son | = 9 years |
| Total | = 24 years |
| Average | = 6 years |

## Annual genetic gain

We have seen that the response to selection in one generation can be estimated based on the selection intensity, heritability of the trait and standard deviation of the trait. By dividing the response to selection by the average generation interval the average annual genetic gain is obtained. (Box, II.3). In the example given in Box II.3 the asymptotic genetic gain per year is estimated. Here the assumption is that generations do not overlap but suceed in tandem. We know that this is not the real life situation and that generations do overlap. In this situation the average annual genetic gain is estimated using the gene flow technique (Hill, 1974). Gene flow calculates genetic response and specifies its dynamics by modelling in detail the flow of genes through the population.

**Box II.3**  Calculation of annual rate of response to selection

---

Assume the selection objective is to improve the birth weight of calves. Average birth weight = 20 kg; SD = 4 kg, and $h^2$ of the trait = 0.3 (assumed). Bulls are selected from the top 5% and female replacements from the best 80% of the herd.

Generation interval between bulls and their progeny = 4 years, and between cows and their progeny= 5 years.

Substituting the appropriate figures in the equation, i for 5% = 2.06 and for 80% = 0.35:

$$\text{Annual R} = [(i_m + i_f) / (L_m + L_f)] \cdot SD \times h^2$$

$$= [(2.06 + 0.35) / (4 + 5)] \cdot 4 \cdot 0.3$$

$$= (2.41/9) \cdot 4 \cdot 0.3 = 0.32$$

Therefore the expected rate of progress will be 0.32 kg/y$^{-1}$.

---

Other factors that affect the genetic progress are as follows:

- Reliability of estimated breeding values
- Selection intensity: The higher the intensity of selection, the higher the response to selection, i.e. the higher the proportion of animals excluded from breeding (or) the smaller the proportion of animals selected, the larger the selection response because of higher selection differential.
- Generation interval: The shorter the generation interval, the more often we can select and the higher our breeding progress in the long run. Generation interval is defined as the average age of the parents at birth of offspring.
- Traits considered: The fewer the traits, the higher the response to selection in the respective traits. If the traits included correlate negatively, then the response to selection will be smaller.
- Variation available: If there is more variation available in the population, the greater the superiority of the selected fraction and therefore the higher the selection response.

## Time factor in selection programme

Interventions in commercial breeding are with the intention of bringing in maximum improvements per year. To obtain maximum returns the genetic gain per generation should be maximised and the average generation interval minimised. In dairy cattle and buffaloes only a few bulls are needed when AI is practised as a means of reproduction. It may be recalled that genes pass from one generation to the next through four paths, viz. from males to males, from males to females, from females to males and from females to females. The selection intensities and generation intervals for each of these paths will be separate. In the absence of selection means such as progeny testing the genetic improvement from bulls producing the next generation males and females ($G_{mm}$ and $G_{mf}$) would be 0. The genetic improvement from cows to cows ($G_{ff}$) will arise because heifer replacements come from selected cows. Genetic gains from cows to bulls ($G_{fm}$) will arise from the fact that only top cows are used as bull dams. The generation interval through the path male to male ($L_{mm}$) and male to female ($L_{mf}$) will be equal to the average number of years the bulls are used for mating when non-progeny tested bulls are used for mating. A comparison of generation interval in different systems is explained in Box II.4. The average age of the cows when their sons and daughters are born (example shown in Box II.5.) will depend on:

- Average age at first calving of the heifers
- Calving interval
- Average number of lactations for which the cows are kept
- Selection intensity

**Box II.4**   Calculation of generation interval in different systems of use of bulls

| System | Adult age (y) | Proving time (y) | Effective use of semen for AI (y) | Effective use of semen for bull production (y) | Av. Age when sons are born (y) | Av. Age when daughters are born (y) | Total |
|---|---|---|---|---|---|---|---|
| Young bulls to breed cows and bulls | 2 | – | 4 | 1 | 2 | (2+3+4+5)/4= 3.5 | 5.5 |
| Young bulls to breed cows & proven bulls to breed bulls | 2 | 8 | 4 | 1 | 8 | (2+3+4+5)/4= 3.5 | 11.5 |
| Proven bulls to breed cows and bulls | | 8 | 3 | 1 | 8 | (8+9+10)/3=9 | 17 |

**Box II.5**   Calculation of average generation interval

| | Age at first calving (y) | Calving interval (y) | No. of lactations | Generation interval | Total generation interval (y) |
|---|---|---|---|---|---|
| Cows to breed bulls | 3 | 1.25 | 5 | (3+4.25+5.5+6.75+8)/5= 5.5 | 9.75 |
| Cows to breed cows | 3 | 1.25 | 3 | (3+4.25+5.5)/3 =4.25 | |

# Breeding value

The individual animal's genetic worth when predicted, based on the difference (deviation) between its performance ($X_i$) and the average performance of its contemporaries ($X_c$) multiplied by the heritability ($h^2$) of the trait, is known as breeding value (BV).

$$BV = X_i - X_c \cdot h^2$$

Breeding value can be estimated based on information on:

- Progeny
- Individual itself
- Pedigree
- Family

## Breeding value based on progeny records

Breeding value is often used for dairy bulls whose genetic merit for milk production is evaluated on the basis of the milk yield of their daughters. When the breeding value is estimated for bulls based on daughter performance, it is only half of the breeding value of the bull and is referred to as the *transmitting ability* of the bull. Accuracy of the BV estimate ($r^2$) increases as the number of progeny increases and is calculated using the formula:

$$r = n/(n+k)$$

Where: $n$ is the number of daughters and k $(4 - h^2) / h^2$

One question usually asked is: how much information does a progeny test give on an animal's breeding value? The question can be reframed as, how many daughters are needed for a bull to have equal information to that given for a cow from her own performance.

We know that accuracy of BV estimation $r^2$ based on the individual's own performance. $r^2 =$, the for progeny test $r^2 = n/(n+k)$

From the above we can derive that: $n = (4 - h^2) / (1 - h^2)$

From the above formula we know that information from five progeny for a bull will provide equal information to that given for a cow from her own performance when the heritability for the trait is 0.25. The higher the heritability, the more daughters needed to get equivalent information. The argument to do progeny testing for cows to increase the accuracy of the breeding value estimation is irrelevant since at least five daughters are required to get reliability equal to that of its own performance when the heritability is 0.25. In the normal course, to get more than five female progeny would be impractical and time consuming. Change in accuracy of the breeding value estimation (heritability is assumed as 0.25. and k (4 − 0.25) / 0.25 = 15) based on progeny tests with a varying number of daughters is given in Figure II.4.

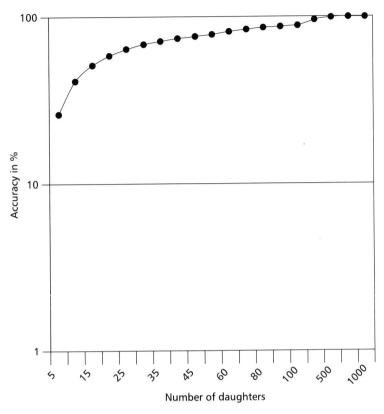

**Figure II.4**  Change in accuracy of the progeny test according to progeny group size when $h^2=0.25$

The BV of the bull based on a progeny test would then be the deviation of the progeny average from the contemporary average multiplied by the factor $2n/(n+k)$ (regression coefficient of the progeny test). The optimum progeny group is determined by the need for a balance between the accuracy of choice and width of choice (Box II.6). It can be seen that the gain in accuracy is rather small after 100 daughters.

Before making any comparison among the daughters it is a prerequisite to make the environmental variations as small as possible. This is accomplished by correcting for the various environmental factors that influence the lactation yield.

## Breeding value from individual records

The breeding value of cows is often estimated using information on individuals. The heritability of a trait increases when several records are compiled on the

**Box II.6** Calculation of breeding value (first standard lactation milk yield, kg) of a bull based on progeny records

| Contemporary average 3500 kg, heritability 0.25 | | | | |
|---|---|---|---|---|
| | Bull 1 | Bull 2 | Bull 3 | Bull 4 |
| Progeny group size | 59 | 85 | 11 | 110 |
| Average milk yield of progeny | 3585 | 4100 | 4250 | 3600 |
| Deviation from contemporary mean | 85 | 600 | 750 | 85 |
| Regression coefficient of the progeny test (n/n+k) | 0.797 | 0.85 | 0.423 | 0.88 |
| Breeding value | 135.6 | 1020 | 634 | 149.6 |

Bull number 2 on account of the higher number of daughters is superior to bull number 3, though the former has a smaller deviation from contemporary mean. Though, bull numbers 1 and 4 have the same deviation from the contemporary mean, bull number 4 is superior because of the higher number of progeny.

animal (example successive lactations). The heritability of average measurements of the same traits on the same animal is obtained according to the formula:

$$h_m^2 = h_1^2 \cdot m / [1 + (m-1)r]$$

where m the number of records; $h_m^2$ the heritability when m records are considered; $h_1^2$ the heritability of a single record; r the repeatability.

The measure of how an animal measured on any one occasion, will repeat its performance during its lifetime is called repeatability. Repeatability of the trait would decide the extent to which accuracy is increased by repeated measurement. It is the correlation between records and is expressed as a proportion from 0.0 (zero repeatability) to 1.0 (complete repeatability).

Results of the equation m/[(1+(m–1)r], solved for 1–6 records with different repeatabilities, are given in Table II.4.

The appropriate factor multiplied by the heritability of the trait gives the regression coefficient for estimating the breeding value. Table II.4 illustrates that while the first one or two extra records increase the accuracy appreciably, further extra records give a diminishing return. It also shows that the advantage of extra records is considerably higher when the repeatability of the trait is low than when it is high. An example to estimate the breeding value based on individual's record is given in Box II.7. ($h^2 = 0.25$; r = 0.4)

With one record cow 1 would have been selected out of the three, cow 2 got an advantage over cow 1 with an additional record for her and cow 3 which gave

**Table II.4**  The equation m/[(1+(m−1)r] solved for 1–6 records with different repeatabilities

| No. of records (m) | Repeatability (r) | | | | | | | | |
|---|---|---|---|---|---|---|---|---|---|
| | 0.1 | 0.2 | 0.3 | 0.4 | 0.5 | 0.6 | 0.7 | 0.8 | 0.9 |
| 1 | 1.00 | 1.00 | 1.00 | 1.00 | 1.00 | 1.00 | 1.00 | 1.00 | 1.00 |
| 2 | 1.82 | 1.67 | 1.54 | 1.43 | 1.33 | 1.25 | 1.18 | 1.11 | 1.05 |
| 3 | 2.50 | 2.14 | 1.88 | 1.67 | 1.50 | 1.36 | 1.25 | 1.15 | 1.07 |
| 4 | 3.08 | 2.50 | 2.11 | 1.82 | 1.60 | 1.43 | 1.29 | 1.18 | 1.08 |
| 5 | 3.57 | 2.78 | 2.27 | 1.92 | 1.67 | 1.47 | 1.32 | 1.19 | 1.09 |
| 6 | 4.00 | 3.00 | 2.40 | 2.00 | 1.71 | 1.50 | 1.33 | 1.20 | 1.09 |

**Box II.7**  Estimation of breeding value for three cows with differing records.

| | Herd average | Individual performance & difference from the herd | | | | | |
|---|---|---|---|---|---|---|---|
| | | Cow 1 | | Cow 2 | | Cow 3 | |
| | | Yield | Difference | Yield | Difference | Yield | Difference |
| Lactation 1 | 3500 kg | 3586 | +86 | 3528 | +28 | 3487 | -13 |
| Lactation 2 | 3890 kg | | | 4114 | +224 | 4114 | +224 |
| Lactation 3 | 4120 kg | | | | | 4154 | +34 |
| Total difference | | | +86 | | +252 | | +245 |
| Average difference | | | +86 | | +126 | | +81.66 |
| $h_m^2$ | | 0.250 | | 0.357 | | 0.417 | |
| Breeding value | | 21.500 | | 44.900 | | 30.100 | |

poor production in the first lactation improved her breeding value with the second and third lactation.

From repeated observations we benefit in terms of extra reliability. However it:

- Involves extra work and extra money
- May delay decisions about which animals to select and
- Could increase the generation interval

Therefore it is essential to strike a balance between the speed of selection and the accuracy of selection. Exercise 7 of the toolbox in the attached CD will help the reader to estimate breeding value of cows.

*Exercise 7.    Breeding value estimation based on individual records.*

## Breeding value based on pedigree information

The use of pedigree information is in principle similar to progeny testing. If the information is limited to sire and dam they can be averaged, being independent. However the issue becomes rather complicated when there is information from several generations (e.g. milk yield of the dam of a young bull as well as progeny test of her father), since these pieces of information are themselves correlated. It can be seen that the correlation between sire's progeny test and son's breeding value is $\frac{1}{2}\sqrt{n/(n + k)}$. Breeding value estimation based on pedigree information is of importance for selection at an age before the individual is old enough to exhibit the trait in itself. (E.g. selecting a young male calf for milk production potential based on dam's performance.) The animal inherits half of its genes from either of the parents; half from the sire and half from the dam. In other words the breeding value of a young male calf estimated based on the dam's performance would be half of that of the dam. Drawing further from Box II.7 the breeding value of the calf from each of these dams is given in Box II.8.

**Box II.8**  Breeding value estimation for male calves born of three cows with differing records

| | Herd average | Individual performance & difference from the herd | | | | | |
| --- | --- | --- | --- | --- | --- | --- | --- |
| | | Cow 1 | | Cow 2 | | Cow 3 | |
| | | Yield | Difference | Yield | Difference | Yield | Difference |
| Lactation 1 | 3500 kg | 3586 | +86 | 3578 | +28 | 3487 | −13 |
| Lactation 2 | 3890 kg | | | 4114 | +224 | 4114 | +224 |
| Lactation 3 | 4120 kg | | | | | 4154 | +34 |
| Total difference | | | +86 | | +252 | | +245 |
| Average difference | | | +86 | | +126 | | +81.66 |
| $h_m^2$ | | 0.250 | | 0.357 | | 0.417 | |
| Breeding value of dams | | 21.5 | | 45.0 | | 34.1 | |
| Breeding value of sons | | 10.8 | | 22.5 | | 17.0 | |

In a similar fashion the breeding value of the calf can also be calculated from the breeding value of his father.

## Breeding value based on half-sibs

Breeding value estimations are possible with information from collateral relatives, mainly full sibs and half-sibs. In dairy cattle while information from half-sibs are readily available should there be a progeny-testing programme, full sib information is not commonly available unless there is embryo transfer programme.

The son's breeding value based on the progeny test of his father is simply the result of the half-sib analysis. All daughters of the father are half sisters of the son. The principle of the progeny test therefore applies to the half-sib analysis. Due to the same reasoning as explained under progeny testing, it is not worth to use half-sib analysis for estimating breeding value of cows (the individual's performance is better in reliability and results are obtained earlier in age). It can be seen that by multiplying the deviation from the contemporary mean with $\sqrt{n/(n+k)}$ the breeding value of a bull can be estimated based on half-sib information. An example given in Box II.9

Breeding value = $\sqrt{n/(n+k)}$ . (half-sib average – contemporary average)

**Box II.9**   Calculation of breeding value (first standard lactation milk yield, kg) of a bull based on half-sib information.

| Contemporary average 3500 kg, heritability 0.25 | | | | |
|---|---|---|---|---|
| | Bull 1 | Bull 2 | Bull 3 | Bull 4 |
| Half-sib group size | 59 | 85 | 11 | 110 |
| Average milk yield of half-sibs | 3585 | 4100 | 4250 | 3585 |
| Deviation from contemporary mean | 85 | 600 | 750 | 85 |
| Regression coefficient of the progeny test $\sqrt{n/(n+k)}$ | 0.893 | 0.922 | 0.650 | 0.938 |
| Breeding value | 75.9 | 553.2 | 487.8 | 79.7 |

Bull number 2 on account of the higher number of half-sibs is superior to bull number 3, though the former has a smaller deviation from contemporary mean. Though bull numbers 1 and 4 have the same deviation from the contemporary mean, bull number 4 is superior on account of the higher number of halb sibs.

Exercise 8 of the toolbox in the attached CD will help the reader to calculate the breeding values of bulls based on the daughters averages.

*Exercise 8.    Estimation of breeding value based on half-sib records.*

# Breeding systems

Selection and breeding are two inseparable aspects of genetic improvement of animals. A breeding system allows the breeder to bring in desired changes in the genetic make-up of a given population. The systems of breeding can be discussed under inbreeding, crossbreeding and the various levels of cross-breeding.

## Inbreeding

### The theory of inbreeding

Inbreeding is the mating between animals that are more closely related than the average relationship of the population. Before discussion further attention is drawn to the examples of inbreeding given in Figure II.5.

All the examples shown in Figure II.5 indicate that the offspring born is through a mating between related individuals; only the extent (intensity) of relationship varies. The common ancestors are shown in shaded boxes. Mating between related individuals can be deliberate or accidental. Many farmers strongly believe that inbreeding will fix good traits of the animals and therefore will improve their productivity. This belief is more wrong than right however, since inbreeding will increase homozygocity among gene pairs that will lead to expression of recessive characters, which are often harmful.

In small populations it is not possible to avoid inbreeding; it can only be minimised or delayed. Restricting the numbers on one or both parents will increase the rate of inbreeding rather steadily and at a faster pace. This often happens when few bulls are used for AI in a small population over a longer period of time. Often several bulls of high genetic merit are imported in developing countries for the AI programme. The genetic base is narrowed by this ending up in a steady accumulation of the bad effects of inbreeding. Bringing in germplasm from many unrelated bulls even at the expense of the quality with respect to its wanted traits is the solution to overcome this situation. Further discussion of this subject given in Part I, Chapter 7.

### Inbreeding coefficient

Coefficient of inbreeding estimates the degree of homozygocity. It is defined as the average percentage of decrease in heterozygocity in the animal to which it applies in relation to the average of animals of the breed at the starting point. The letter F denotes the inbreeding coefficient. From the definition it can be seen that the inbreeding coefficient is a measure of relative and not of absolute decrease in heterozygocity.

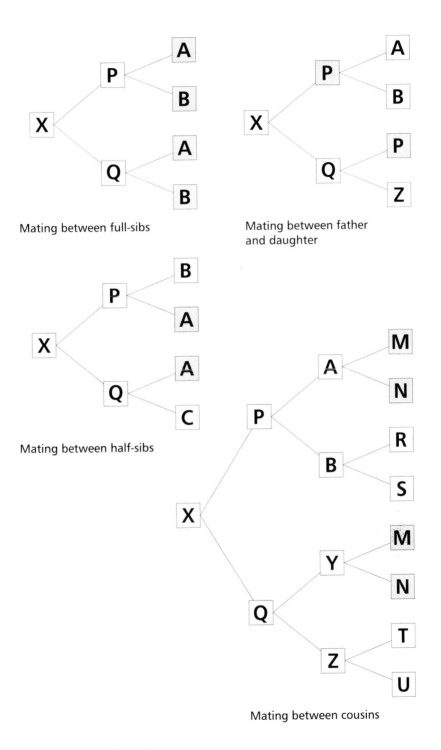

Mating between full-sibs

Mating between father and daughter

Mating between half-sibs

Mating between cousins

**Figure II.5**  Examples of inbreeding

$$\text{Coefficient of inbreeding: } F_x = \Sigma[(\tfrac{1}{2})^{n+n'+1}(1+F_a)]$$

where: $\Sigma$ means the sum of all common ancestors;
$n$ is the number of generations from sire to the common ancestor;
$n'$ is the number of generations from dam to the common ancestor;
$F_a$ is the inbreeding coefficient of the common ancestor (if this is not known, it is taken as zero).

To calculate the inbreeding coefficient one should know the pedigree of an animal. The extent of inbreeding for an animal will be rather small when its ancestors are common only in the third and above generations.

If the animal is inbred, this means that its parents are related. This relationship can be expressed as the coefficient of relationship and is double the inbreeding coefficient. The coefficient of relationship measures the similarity of genotypes.

$$\text{Coefficient of relationship between X and Y: } a_{xy} = \Sigma(\tfrac{1}{2})^{n+n'}$$

where: $n$ is the number of generations from the common ancestor to one animal;
$n'$ is the number of generation of the common ancestor to the other animal;
$F_a$ is the inbreeding coefficient of the common ancestor (if this is not known, its is taken as zero).

The coefficient of relationship between:

$$\text{Parent and offspring} = (\tfrac{1}{2})^1 = 0.50$$
$$\text{Half-sibs} = (\tfrac{1}{2})^{1+1} + (1+0) = 0.25$$
$$\text{Full sibs} = \Sigma[(\tfrac{1}{2})^{1+1}(1+0) + (\tfrac{1}{2})^{1+1}(1+0)] = 0.50$$

The coefficient of relationship is often calculated before arranging nominated mating in nucleus herds to control the level of inbreeding in the progeny generation.

## Rate of inbreeding

Inbreeding can occur by chance in relatively small populations in situations wherein it is not deliberately avoided. The rate of inbreeding that can happen per generation in such situations is calculated using the formula:

$$\text{Rate of inbreeding / generation} = 1/(8N_m) + 1/(8N_f)$$

where: $N_m$ is the number of males used as parents and
$N_f$ the number of females used as parents.

In cattle populations the increase in rate of inbreeding is mainly the result of using fewer bulls for breeding. An example is given in Box II.10.

**Box II.10** Calculation of rate of inbreeding

| Cows | Bulls used | $1/(8N_m)$ | $1/(8N_f)$ | Rate of inbreeding | |
|------|-----------|-----------|-----------|------------|--------|
| | | | | Proportion | In % |
| 500 | 5 | 0.025 | 0.00025 | 0.02525 | 2.525 |
| 1000 | 5 | 0.025 | 0.000125 | 0.02513 | 2.513 |
| 10,000 | 2 | 0.0625 | 0.000013 | 0.062513 | 6.251 |
| 50,000 | 2 | 0.0625 | 0.000003 | 0.062503 | 6.250 |
| 10,000 | 10 | 0.0125 | 0.000013 | 0.01251 | 1.251 |

From the example shown in Box II.10 we realise that most inbreeding occurs as a result of using too few bulls. The role of females in changing the rate of inbreeding is rather trivial. This emphasises the point that in AI programmes for cattle the number of bulls used shall not be too small. The rate of inbreeding in a population of 10,000 cows was reduced from 6.251% to 1.251% by increasing the number of bulls from 2 to 10. The use of a sufficiently large number of sires is even more important in crossbreeding programmes targeted towards new breed formation.

### Effects of inbreeding

To explain the effects of inbreeding, a simple situation of the action of one pair of genes would be useful. Animals that are Aa at a particular locus produce gametes A and a in equal proportion. The possible combinations in the next generation are AA, Aa and aa in the ratio 1:2:1. The single gene action example can be extrapolated to a multigene action situation and the following general statements made:

- Mating between heterozygous individuals leads to a dispersal of genotypes and will increase the variability.
- Mating between homozygous individuals will alternatively result in genetic uniformity.

It may be noted that inbreeding is analogous to this simplified system. Mating between closely related individuals would tend to increase the proportion of homozygotes. Most of the lethal and harmful genes are recessive in nature and by inbreeding there is increased chance for these genotypes to exhibit. This is one reason for what is referred to as inbreeding depression. Another reason is the proportionate reduction in heterozygocity and the resultant loss of benefits associated with it. Inbreeding is stated to be the opposite of crossbreeding and the inbreeding depression the converse of heterosis. For characters that benefit most by crossbreeding, inbreeding depression is greatest. Traits such as growth, milk yield and reproduction are stated to be affected by inbreeding depression.

Deliberate inbreeding is attempted in species where the reproductive rate is very high (example poultry) to produce highly inbred lines with the idea of later crossing between the inbred lines to harvest the heterosis effect. This system however is not practical in animals with a low reproduction rate such as cattle and buffaloes because the inbreeding process leads to too great loss of lines and too great loss in performance to be sustainable. However, this system may be beneficial provided new techniques in reproductive physiology enabling harvest of hundreds of eggs from each cow and its *in-vitro* fertilisation emerge.

In a selection programme the objective is to use the best of the available germplasm to produce the next generation and mating between related animals is unavoidable. But there has to be a compromise between the increased progress achieved by increasing the selection intensity and the ill effects resulting from inbreeding depression.

To minimise and delay inbreeding the following steps are recommended:

- While bringing in germplasm from outside the population as bulls, semen, cows, or embryos ensure that it is not closely related
- Arrange to procure many individuals
- Practise a bull rotation programme to change bulls from a population by the time their own daughters are ready for mating
- Practise grouping bulls into families so that all related bulls are maintained as one family and the family is rotated between zones of AI
- Nominated mating shall be followed in the nucleus herd (from which bulls are selected) to produce offspring that are not related
- If considerable inbreeding has arisen in a herd it shall be countered by the use of sires from an unrelated population

## Crossbreeding

### Principle

One of the most rapid ways of making a change in gene frequency within a given population is by bringing in new genes from outside sources. The introduction of genotypes from other breed/breeds into the indigenous breed or a specified group of animals (it can also be animals that belong to no specified breed) is referred to as crossbreeding. Even the most suited breed to a given agroeconomic and sociocultural situation can benefit by crossbreeding by way of productivity enhancement and better use of the improved feeding and management provided. In the tropics crossbreeding with exotic dairy breeds is widely used for productivity enhancement of dairy cattle.

We know that the effects of a large number of genes can be:

- **Additive**–heterozygotes are halfway in performance between homozygotic pairs
- **Dominant**–one of the alleles can be totally or partially dominant in effects over the other, i.e. the performance of heterozy gotes will be the same or closer to the dominant homozy gotes
- **Epistatic**–the genotype at one locus influences the gene action at another locus

Most of the breeds in the world have been developed through systematic and continuous selection and breeding over a long period of time. It is most likely that a good proportion of the genes carried by these breeds are possibly homozygous for different alleles. The difference in gene frequency between the two breeds might have occurred because the breeds either had a different selection history or because the gene frequencies diverged by chance. Crossing between such breeds would naturally bring in a higher degree of heterozygocity in the crossbred. The more distant the breeds crossed, the higher the heterozygocity. Thus a cross between *Bos taurus* and *Bos indicus* would have higher heterozygous alleles than a cross between two *Bos taurus* breeds.

## Heterosis

The expected productivity of a crossbred by additive gene action would be the average productivity of the two parental breeds, the productivity recorded under the same conditions of feeding and management. However, it is often observed that the productivity of the crossbreds is higher than the average of their parents. This additional production is called hybrid vigour or heterosis. Heterosis is measured as a deviation of the performance of the crossbred from the average of its two parental breeds. The dominance gene action at the heterozygous loci mainly contributes to heterosis. As such the occurrence of heterosis is directly proportional to the degree of heterozygocity. $F_1$ crosses express maximum heterosis (100%) which is halved at each of the forward crossings. In situations where the productivity bulls and/or semen brought in from outside sources are not available, a satisfactory way to estimate the extent of heterosis is to compare the performance of the $F_1$ and the $F_2$ at the same time and under the same conditions. We know that they are equal in terms of genetic make-up; but differ in proportion of genes present in the heterozygous state. So while $F_1$ expresses all the heterosis (100%), $F_2$ would express only half (50%). The example given in Box II.11 describes the steps to calculate heterosis.

**Box II.11**   Calculating heterosis from records of F$_1$ and F$_2$

Average first standard milk yield of Jersey × Local

F$_1$ = 2400 kg

F$_2$ = 2200 kg

Difference between the two groups; 2440 – 2200 = 200 kg = 50% heterosis

Therefore estimated 100% heterosis is = 200 · 2 = 400 kg

Knowledge about the amount of heterosis is important for taking decisions on the extent of crossbreeding one has to adapt in a given situation. If heterosis is significant it is advantageous to continuously produce F$_1$ generations; if it is negligible, interbreeding the crosses is the best option.

To have a continuous production of F$_1$ animals, it is necessary to keep suffi-cient stock of pure indigenous population to produce crossbreds as well as to maintain the population. F$_1$ females cannot be allowed to produce replacement females. Given the need to keep more than two-thirds of the population as indigenous stock, the overall average production in the population will not be commensurate with the benefits accrued through heterosis.

Considering the above, it is advisable to go in for permanent recreation of F$_1$ animals in crossbreeding programmes of dairy cattle. It is also argued that if heterosis is very significant, there is not much genetic advantage by simply increasing the exotic blood level.

Crossbreeding is widely used for improving milk production in dairy cattle of the tropics. Crossbreeding results in India (Taneja and Chawla, 1978; Nair, 1973) have shown that there was considerable increase in milk production through crossbreeding of the local zebu cattle with exotic dairy breeds. The study of Cunningham and Syrstad (1987) based on many crossbreeding experiments in the developing world, showed that there was increase in milk yield of the crossbreds with exotic inheritance up to 50% and thereafter additions of exotic blood led to little additional improvement (Table II.5)

**Table II.5**   Milk yield of cows in tropics with different levels of exotic inheritance

| Exotic blood % | 0 | 12.5 | 25 | 37.5 | 50(F$_1$) | 62.5 | 75 | 87.5 | 100 | 50(F$_2$) |
|---|---|---|---|---|---|---|---|---|---|---|
| Milk yield kg | 1052 | 1371 | 1310 | 1553 | 2039 | 1984 | 2091 | 2086 | 2162 | 1523 |
| SE (±) | 39 | 170 | 158 | 100 | 28 | 75 | 45 | 84 | 50 | 92 |

*Source: Cunningham and Syrstad (1987)*

Another finding from the above study was the comparatively low yield of the $F_2$ generation. This could be the result of loss of heterosis by half, which was not compensated by a good selection programme. Increasing the level of exotic inheritance among crossbreds in the tropics resulted in reduced survivability (Vaccaro, 1979).

### Upgrading

Continuous backcrossing of the indigenous stock with the donor breed is called upgrading. It can be seen that the level of inheritance of the donor breed in the indigenous population gradually increases through generations of backcrossing. The level increases from 50% to 98.4% in six generations.

## New breed formation

Many crossbreeding programmes in low-input systems aim at maintaining the positive qualities of the indigenous population such as hardiness, adaptability and disease resistance, with an addition of economic qualities (example milk yield, reproductive efficiency) of the donor breed / breeds. This is possible in two stages:

- Reaching the desired level of donor inheritance by crossbreeding
- Genetic improvement of the crossbred without altering the blood level through selection.

Such new breeds are called synthetic breeds. Synthetic breeds can be developed from a combination of two or more breeds and with varying proportions of inheritance from each of the participating breeds. In large-scale crossbreeding programmes synthetic breeds are acquiring the blood of more than one donor breed mainly due to changing policies over a period of time. Synthetic breed formation has considerable advantages over permanent crossbreeding systems in terms of operational simplicity. Once the desired blood level is reached, the synthetic breed can be treated as 'pure breed in the making' and the breeding programme will be one of selective breeding. It may be recalled that many of the established breeds available now were at one point of time synthetic in the sense that they were evolved from a stock of crossbreds. A large number of breeds with national identities were evolved from common foundations (example British Friesians, Canadian Holsteins, Israeli Holsteins). In each of these countries the foundation stock were selected with different aims and that made differences among these breeds. Some of the established synthetic dairy cattle breeds are shown in Box II.12.

The simplest synthetic breed is the one developed out of $F_1$ animals through inter-se mating for several generations and applying systematic selection at every generation. Here the percentage of exotic inheritance would be 50 from

**Box II.12** Some examples of synthetic breeds

| Breed | Composition | Origin |
|---|---|---|
| Australian Milking Zebu | 0.33 Sahiwal + Red Sindhi and 0.67 Jersey | Australia |
| Jamaica Hope | 0.8 Jersey, 0.05 HF and 0.15 Sahiwal | Jamaica |
| Sunandini | 0.5 Local ND and 0.5 Jersey, HF and Brown Swiss | Kerala, India |
| Karan Swiss | Brown Swiss and Sahiwal | Haryana, India |

each of the breeds, the indigenous and the donor. In this system 100% of possible heterozygocity is obtained. With the involvement of more than two breeds, the expected performance level can be calculated as a function of the expected heterosis and additive effects. However, the production level of a three-breed synthetic will be lower than the average of a three-breed rotational cross. Synthetic breeds have the advantage over rotational crossing in more uniform performance and the possibility of exercising selection.

# B. Statistics

*The statistical basics described in this chapter follow a classical structure, often used in livestock science textbooks. It starts with the definition of the population, sample and sampling methodologies. The means of variability and normal distribution are explained and supported by examples and case studies. The methodology of hypothesis testing is followed by an introduction into regression and correlation theories. This chapter will assist the reader to find the statistical and mathematical roots of the calculations used in the genetics part of this book, before referring to more basic literature.*

# Population and sample

## General

Statistics can be regarded as a body of methods used to summarise data (descriptive statistics) and to draw conclusions from the data (statistical inference). It is essential to familiarise biologists with statistical methods, in particular those applied in biological sciences (often called biostatistics).

Population is the set of things one wishes to study and should have at least one attribute in common (e.g. cattle, beef breeds, Friesian cows in UK). In practice one seldom, if ever, works on a population due to reasons of

- Impracticability
- Financial limitations
- Time limitations

Analyses are usually based on samples drawn from the population to be studied for the purpose of drawing inferences on the population. For example, to obtain an estimate of the first lactation milk yield of Holstein Friesian (HF) cattle in a certain country, we record the lactation yield of a sample of first calvers in different parts of the country. From their performance we make inferences about the first lactation milk yield of the population of HF cows in the country. Estimating the yield taking samples from high feeding farms (the production level is expected to be higher), will overestimate the true performance because the sample does not represent the population. To get an unbiased estimate of the population samples should be drawn from all systems of feeding and management. The method of drawing samples which facilitates representation to all situations, is called simple random sampling. In simple random sampling equal chance is given to all rearing modes of the population to be represented. Such a sample can be drawn from a population by use of a table of random digits, draw of cards from a deck of shuffled cards without looking at the face, etc.

How big should the sample be to secure meaningful results? An unnecessarily large sample wastes time and money; too small a sample yields meaningless results.

There are two types of attributes (observations), discrete and continuous. Attributes such as height, weight, milk yield, age at calving etc. are from a continuous scale; even if two observations very close by are picked one can still imagine another in between the two. Attributes such as sex or number of calves born are not on a continuous scale; they are referred to as discrete. The methods described in the following sections are mainly for continuous variables.

# Describing a sample

For meaningful inferences it is necessary to simplify the dataset through various procedures such as rearrangements, tabulations, graphs and summarising statistics. The various ways of describing data are illustrated using data on the birth weights of 181 Sunandini male calves given in Table II. 6.

**Table II.6**  Birth weight (kg) of Sunandini male calves

| | | | | | | | | | | | | |
|---|---|---|---|---|---|---|---|---|---|---|---|---|
| 30.4 | 30.9 | 30.3 | 27.9 | 28.3 | 28.7 | 29.3 | 29.8 | 30.2 | 30.7 | 31.3 | 31.9 | 32.8 |
| 23.6 | 26.3 | 27.2 | 27.9 | 28.4 | 28.8 | 29.3 | 29.8 | 30.2 | 30.8 | 31.4 | 32 | 32.9 |
| 24.1 | 31.7 | 32.2 | 27.9 | 29.6 | 28.8 | 29.3 | 33.6 | 30.3 | 30.8 | 31.4 | 28.5 | 33 |
| 24.3 | 26.5 | 27.3 | 28 | 32.1 | 28.9 | 29.4 | 29.8 | 26.7 | 30.8 | 31.5 | 32 | 33.3 |
| 31.8 | 32.5 | 34.6 | 33.5 | 28.5 | 29 | 26.7 | 29.9 | 30.3 | 26.9 | 31.6 | 28 | 29.5 |
| 25 | 29.6 | 27.4 | 30.8 | 32.2 | 29 | 29.4 | 29.9 | 30.4 | 30.9 | 31.7 | 32.1 | 29.9 |
| 25.1 | 34.7 | 29.8 | 27.4 | 28.5 | 29 | 26 | 27 | 26.3 | 27.3 | 25.3 | 30.5 | 30 |
| 28.6 | 33.8 | 27.5 | 28.1 | 28.6 | 29 | 29.5 | 31.9 | 30.5 | 30.9 | 31.7 | 28 | 33.8 |
| 25.5 | 26.8 | 27.6 | 28.1 | 27.8 | 29.1 | 26.7 | 30 | 30.5 | 31 | 31.7 | 32.4 | 33.9 |
| 30.1 | 30.6 | 31.1 | 28.2 | 28.6 | 29.1 | 29.6 | 30 | 30.5 | 31 | 31.8 | 30.1 | 28.4 |
| 25.9 | 26.9 | 27.7 | 28.2 | 28.6 | 29.2 | 24.7 | 26.6 | 27.3 | 31.1 | 25.7 | 28.4 | 27.6 |
| 32.5 | 34.3 | 32 | 28.3 | 28.6 | 29.2 | 29.7 | 22.9 | 26.1 | 27.2 | 31.8 | 32.6 | 36.4 |
| 26.1 | 27 | 27.8 | 28.3 | 28.7 | 29.2 | 29.7 | 30.1 | 30.6 | 31.1 | 28 | 32.7 | 29.4 |
| 26.1 | 27 | 27.8 | 28.3 | 28.7 | 29.2 | 29.8 | 30.2 | 30.7 | 31.2 | 31.9 | 32.7 | |

## Array

For a better understanding of the dataset it can be rearranged as an array. An array is the arrangement of the observations according to size. The birth weight data of Table II.6 is rearranged into an array shown in Table II.7.

From the array we can notice that
- All the observations lie between 22.9 and 36.4
- 49% are between 28 and 30.9

## Frequency table

The data are still cumbersome and can be further refined as a frequency table. In a frequency table the observations are grouped into different class intervals. The number of observations in each class is called frequency. The class interval can be chosen arbitrarily. The data on birth weight converted into a frequency table with a class interval of one kg are given in Table II 8.

**Table II.7** Birth weight (kg) of Sunandini male calves, depicted as an array

| | | | | | | | | | | | | |
|---|---|---|---|---|---|---|---|---|---|---|---|---|
| 22.9 | 26.1 | 27.2 | 27.9 | 28.3 | 28.7 | 29.3 | 29.8 | 30.2 | 30.7 | 31.3 | 31.9 | 32.8 |
| 23.6 | 26.3 | 27.2 | 27.9 | 28.4 | 28.8 | 29.3 | 29.8 | 30.2 | 30.8 | 31.4 | 32 | 32.9 |
| 24.1 | 26.3 | 27.3 | 27.9 | 28.4 | 28.8 | 29.3 | 29.8 | 30.3 | 30.8 | 31.4 | 32 | 33 |
| 24.3 | 26.5 | 27.3 | 28 | 28.4 | 28.9 | 29.4 | 29.8 | 30.3 | 30.8 | 31.5 | 32 | 33.3 |
| 24.7 | 26.6 | 27.3 | 28 | 28.5 | 29 | 29.4 | 29.9 | 30.3 | 30.8 | 31.6 | 32.1 | 33.5 |
| 25 | 26.7 | 27.4 | 28 | 28.5 | 29 | 29.4 | 29.9 | 30.4 | 30.9 | 31.7 | 32.1 | 33.6 |
| 25.1 | 26.7 | 27.4 | 28 | 28.5 | 29 | 29.5 | 29.9 | 30.4 | 30.9 | 31.7 | 32.2 | 33.8 |
| 25.3 | 26.7 | 27.5 | 28.1 | 28.6 | 29 | 29.5 | 30 | 30.5 | 30.9 | 31.7 | 32.2 | 33.8 |
| 25.5 | 26.8 | 27.6 | 28.1 | 28.6 | 29.1 | 29.6 | 30 | 30.5 | 31 | 31.7 | 32.4 | 33.9 |
| 25.7 | 26.9 | 27.6 | 28.2 | 28.6 | 29.1 | 29.6 | 30 | 30.5 | 31 | 31.8 | 32.5 | 34.3 |
| 25.9 | 26.9 | 27.7 | 28.2 | 28.6 | 29.2 | 29.6 | 30.1 | 30.5 | 31.1 | 31.8 | 32.5 | 34.6 |
| 26 | 27 | 27.8 | 28.3 | 28.6 | 29.2 | 29.7 | 30.1 | 30.6 | 31.1 | 31.8 | 32.6 | 34.7 |
| 26.1 | 27 | 27.8 | 28.3 | 28.7 | 29.2 | 29.7 | 30.1 | 30.6 | 31.1 | 31.9 | 32.7 | 36.4 |
| 26.1 | 27 | 27.8 | 28.3 | 28.7 | 29.2 | 29.8 | 30.2 | 30.7 | 31.2 | 31.9 | 32.7 | |

**Table II.8** Birth weight (kg) of Sunandini male calves as a frequency table (class interval one kg)

| Class no. | 1 | 2 | 3 | 4 | 5 | 6 | 7 | 8 | 9 | 10 | 11 | 12 | 13 |
|---|---|---|---|---|---|---|---|---|---|---|---|---|---|
| Interval | <24 | 24–24.9 | 25–25.9 | 26–26.9 | 27–27.9 | 28–28.9 | 29–29.9 | 30–30.9 | 31–31.9 | 32–32.9 | 33–33.9 | 34–34.9 | 35.8 &> |
| Frequency | 2 | 3 | 6 | 14 | 20 | 29 | 31 | 29 | 21 | 15 | 7 | 3 | 1 |

It is apparent that the same dataset can be put into different frequency tables by choosing different class intervals and/or by changing the starting point of the class interval. The decision to fix the class interval is on the experimenter based on his requirement. However, before deciding on the class interval the range of the dataset has to be taken into consideration.

**Arithmetic mean**

Arithmetic mean is the number most commonly used to describe the centre of the distribution. Arithmetic mean or simply mean is defined as the sum of all observations ($\Sigma X$) divided by the number ($n$) of observations. The symbol

$\mu$ (mu) is used to denote the arithmetic mean of the population and $\overline{X}$ for symbolising the sample mean.

The formula for the sample mean is $\overline{X} = (X_1 + X_2 + \ldots + X_n)/n$. The sample mean of the 181 records on birth weights of Sunandini male calves given in Table II.6 is:

$$30.4 + \ldots\ldots\ldots\ldots\ldots\ldots\ldots\ldots + 29.4 = 5334.8$$

$$n = 181$$

$$\overline{X} = (5334.8 \div 181) = 29.47$$

There are situations wherein means have to be calculated from averages of groups of observations. When the number of observations in each group differs, the number of observations in each group should be taken into account while calculating the overall mean. This is called weighted mean. The example given in Box II.13 illustrates the procedure for calculating the weighted mean.

**Box II.13**    Procedure to calculate weighted mean

| Herd | Number of observations | Average lactation yield (kg) |
|------|------------------------|------------------------------|
| A | 123 | 8970 |
| B | 45 | 11,245 |
| C | 453 | 7532 |

The average lactation yield is:
$[8970 \cdot 123 + 11245 \cdot 45 + 7532 \cdot 453] / (123 + 45 + 453) = 8085.9$ calculated from the real observations. It is not the average of the three averages $(8970 + 11245 + 7532) \div 3 = 9249$.

The **median** is the middle number in an array of data $= (n + 1) \div 2$ where n is the number of observations. When the number of observation is odd it is the middle number and when the number of observations is even it is the arithmetic mean of the two middle numbers.

# Measure of variability

## Introduction

**Box II.14**   Example showing first lactation yield (kg) of two herds with equal arithmetic means but different spread

| Dataset observations | Formula | Mean |
|---|---|---|
| 8524, 4445,10,313, 3467, 9875, 7546 | $\Sigma X/n$= 44,170/6 | 7361.7 |
| 7869, 8234, 6918, 6456, 6875, 7818 | $\Sigma X/n$= 44,170/6 | 7361.7 |

The two datasets in Box II.14 have the same mean whereas the first set is more variable than the second. In this situation, mean alone does not describe the data satisfactorily. We need an additional parameter to describe the two sets more precisely. Measuring the variability of the data does this. Variability relates to the spread of the data and can be measured in different ways.

### Range of the data

Range is the difference between the smallest and the largest observation in the dataset. In set 1 the range is the difference between 10,313 and 3476 kg, and in set 2 it is the difference between 8234 and 6456 kg (Box II.15).

**Box II.15**   Equal arithmetic means with different range

| Dataset | Formula | Mean | Range |
|---|---|---|---|
| 8524, 4445,10,313, 3467, 9877, 7546 | $\Sigma X/n$ = 44,172/6 | 7362 | 10,313 – 3476 = 6837 |
| 7869, 8234, 6918, 6456, 6877, 7818 | $\Sigma X/n$ = 44,172/6 | 7362 | 8234 – 6456 = 1778 |

## Mean absolute deviation

Mean absolute deviation of the observations tells how far 'on average' the individual observations are from the arithmetic mean of the set of data. The calculation of mean deviation from the dataset 1 used earlier is given in Box II.16

From the example in Box II.16 it may be seen that dataset 1 has a greater mean deviation (2270.67) than set 2 (611.67). The mean deviation though seldom used in statistics, is a useful step in describing the principle of measuring the variability precisely.

**Box II.16**  Calculation of mean deviation

| Number of observations = 6 | Set 1 | | Set2 | |
|---|---|---|---|---|
| | Observation | Deviation from mean | Observation | Deviation from mean |
| | 8524 | – 1162 | 7869 | – 507 |
| | 4445 | 2917 | 8234 | – 872 |
| | 10313 | 2951 | 6918 | 444 |
| Mean = 7362 | 3467 | 3895 | 6456, | 906 |
| | 9877 | – 2515 | 6877 | 485 |
| | 7546 | – 184 | 7818 | – 456 |
| ΣDeviation with sign | | 0 | | 0 |
| ΣDeviation without sign | | 13624 | | 3670 |
| Mean deviation without sign | | 2270.67 | | 611.67 |

## Mean square deviation (Variance)

Earlier we found that the sum of the mean deviations is zero. In order to overcome this problem, the deviations of individual observations from the mean are squared. By squaring, the negative figures become positive and the problem of getting a sum of zero will be resolved. The mean squares deviation is known as variance and is a very important attribute in quantitative genetics.

The formula for calculating the variance of a sample is $\Sigma(X - \overline{X})^2/(n - 1)$. The denominator is $n - 1$ since the variance of the sample and not of the population is normally calculated. It can be shown algebraically that (see Box II.17)

$$\Sigma(X - \overline{X})^2/(n-1) = \{\Sigma X^2 - (\Sigma X)^2/n\}/(n-1)$$

where: $\Sigma X^2$ to the sum of the squared observations: all the observations are squared first and added.
$(\Sigma X)^2$ to the square of the sum of the observations: all the observations are added and the square then calculated.

The algebraically simpler formula, $\{\Sigma X^2 - (\Sigma X)^2/n\}/(n - 1)$ is commonly used because only three attributes are required in this case.
The sign $\sigma^2$ (sigma squared) is used to symbolise population variance. The sample variance is denoted by $s^2$.

**Box II.17**  Algebraic derivation of alternative formula to calculate the variance of the sample

$$\Sigma(X - \overline{X})^2 = \Sigma(X^2 - 2X\overline{X} + \overline{X}^2)$$
$$= \Sigma X^2 - 2\overline{X} \cdot \Sigma X + n\overline{X}^2$$
$$= \Sigma X^2 - 2\Sigma X/n \cdot \Sigma X + n\Sigma X/n \cdot \Sigma X/n$$
$$= \Sigma X^2 - 2 \cdot (\Sigma X/n) \cdot \Sigma X + (\Sigma X \cdot \Sigma X)/n$$
$$= \Sigma X^2 - (\Sigma X)^2/n$$

When calculating the variance of the sample, $(n-1)$ is used in place of n. The '−1' part takes care of the loss of one degree of freedom while sampling. If you are given the mean of four observations you can write all the observations except one. The last observation is fixed. In other words you have freedom to write only $n-1$ number of observations in the case of a sample.

A fundamental property of variance is that of additiveness whereby the separate causes of variation can be added or subtracted. The variance of two means of two groups of observations is half the variance of all observations. Also the variance of the means of groups, each consisting of n observations is equal to variance divided by n. If there is a distribution of observations with variance $\sigma^2$, and then many samples of size n are taken from that distribution, the mean of the samples will themselves be distributed with a variance of $\sigma^2/n$ (original variance). The greater the value of $n$, the smaller the variance of the sample means.

## Standard deviation

Variance is the precise term describing variability and the units of variance are squared. The squaring of the deviations from the mean was done for the purpose of avoiding a zero result. It is only logical then to bring the result to the same scale as the original observations and for this its square root is obtained. The square root of variance is called standard deviation. For descriptive purposes standard deviation is the most commonly used expression. The standard deviation is in the same scale as the original deviation. The formula for sample standard deviation ($s$) is:

$$s = \sqrt{\{\Sigma X^2 - (\Sigma X)^2/n\}/(n-1)}$$

## Coefficient of variation

The ratio of the standard deviation to the mean is referred as the coefficient of variation (CV)

$$CV = s / \overline{X}$$

The coefficient of variation is expressed as percentage or as proportion. CV is a useful attribute for comparing the standard deviations of samples with different means. See Box II.18 for an example.

**Box II.18** First standard lactation yield of cows in different regions

| Cow type | Mean | Standard deviation | CV (%) |
|---|---|---|---|
| HF in Israel | 9800 | 750 | 7.65 |
| HF in India | 3500 | 600 | 17.14 |
| HF x Local in India | 2100 | 550 | 26.19 |
| Local non-descript | 400 | 100 | 25.00 |

By considering the standard deviation one may infer that the variation is greatest in the HF cows of Israel, followed in order by HF in India, HF x local crossbreds in India and local non-descript. When the standard deviation is standardised to the mean the picture is almost the reverse.

To compare the variability of samples with different means, the CV is more appropriate than the standard deviation.

## Standard error

While variability of a set of observations is described using standard deviation, the spread of several means drawn from a population is described by standard error (SE). If the original population variance is $\sigma^2$, the variance of the sample mean can be calculated as $\sigma^2/n$. The square root of the variance of the sample mean (i.e., standard deviation of the mean) is the standard error.

$$SE = \sqrt{\sigma^2/n} = \sigma/\sqrt{n}$$

We may not normally know $\sigma^2$, but have calculated $s^2$; thus we can calculate

$$SE = \sqrt{s^2/n} = s/\sqrt{n}$$

The standard error enables an indication of how good (or poor) sample mean may be as an estimate of population mean.

# Normal distribution

When a large number of observations taken from a population with regard to a particular trait (example birth weight of Sunandini male calves) are grouped into a frequency table, it can be seen that observations are distributed around the mean with most actually around the mean and a few in the frequency classes on either side. The distribution of data on birth weight of Sunandini male calves (from frequency Table II.8) plotted in Figure II.6. Please note that it has the shape of a bell.

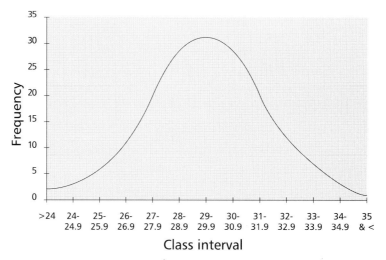

**Figure II.6**   Frequency distribution of data on birth weight of Sunandini male calves plotted as a histogram with a bell-shaped curve superimposed

This type of distribution is called normal distribution and the bell-shaped curve is called the normal curve. The normal distribution is completely defined by the mean and its standard deviation and there will be different normal curves for different means and standard deviations. By knowing the mean and standard deviation it is possible to draw the normal curve.

The proportion of observations falling between two numbers will be equal to the area above the horizontal axis under the normal curve between these two numbers. In a normally distributed set of observations called Xs, with mean $\mu$ and standard deviation s, $(X-\mu)/\sigma$ will normally be distributed with mean 0 and standard deviation 1. $(X-\mu)/\sigma$ is referred to as standard normal variate and is given the symbol $z$. The area under its distribution curve is tabulated and given in Annex 1. The columns are labelled $z$ and $\lambda$: $\lambda$ is the proportion of the area under the standard normal curve to the left of $z$. For example, for $z = 1.50$, $\lambda = .9332$: Thus, 93.32 % of the area lies below $z = 1.50$. An example is given in Box II.19.

If we know the original variance (s²) we can calculate variance of the sample mean as $s^2/n$, the square root of which will be equal to $s/\sqrt{n}$ , called standard error. Since the distribution of these means approximates to normal, we can say that 95% of all the sample means will lie within the range of ±2 SE (standard error) from the mean. This range is known as 95% confidence interval of the mean while the end points are known as 95% confidence limits of the mean. Here what we are saying is that if there are 100 such confidence intervals 95 of them would include the population mean.

**Box II.19** Calculation of confidence interval from lactation yield of HF cows in two countries

From Table II.6 we can calculate the mean birth weight of a sample set of 181 Sunandini male calves as 29.47 kg with a standard deviation of 2.37 kg. We are interested in knowing the average birth weight of the Sunandini male calves in general or the population mean, (μ). Though the best estimate of μ is the sample mean, it is certain that μ is not exactly 29.47 kg (sample mean). But it is somewhere around the sample mean. Let us assume that the standard deviation of the population (birth weight of Sunandini male calves) is the sample standard deviation, 2.37 kg. We can calculate the standard deviation of samples of size 181 as $2.37/\sqrt{181}$ = 0.176. To write down a 95% confidence interval for the μ, we find from Annex I that the distance on each side of μ within which 95% of the sample means lie is 1.96 standard deviation of the sample. So the population mean would lie between 29.47 – (1.96 · 0.176) and 29.47 +(1.96 · 0.176) = 29.13 and 29.81 kg. This interval, called confidence interval, for a parameter is an interval obtained from a sample by some specified method such as that in repeated sampling; 95% of the intervals thus obtained includes the value of the parameter. Confidence intervals can also be set for other limits.

Two important points to be learned from the example in Box II.19 are:

- To find the proportion of observations lying above or below one particular observation, look into the z column of Annex 1 for a number equal to (observation – average) / standard deviation and find the λ value against it.
- To find the top or bottom proportions (percentages) of the observations, look into λ column of Annex 1 for a number equal to the desired proportion, get the corresponding z value and multiply this value by the standard deviation to get the upper or lower limit as the case may be.

This principle is employed in animal breeding to find out the cut-off limits for selecting animals.

A normal distribution can be regarded as a frequency distribution with some well-defined characteristics, determined by mean and variance. It has a bell-shaped appearance that can be broader or narrower. It can be seen from Annex 1 that:

| | | |
|---|---|---|
| $\mu \pm 1.00\ \sigma$ | contains | 68.26% of all observations |
| $\mu \pm 1.96\ \sigma$ | contains | 95.00% of all observations |
| $\mu \pm 2.00\ \sigma$ | contains | 95.44% of all observations |
| $\mu \pm 2.50\ \sigma$ | contains | 98.76% of all observations |
| $\mu \pm 3.00\ \sigma$ | contains | 99.74% of all observations |

It is also implied that the remaining observations lie on either side of the normal curve. In the first case 15.86% [(100 − 68.26)/2] = (31.74/2) of the observations are above the mean plus one standard deviation and 15.86% below the mean minus one standard deviation.

# Hypothesis testing

A hypothesis is a preconceived idea about what is going to happen. Suppose we know the population mean and variance. We draw a sample of observations from this population and calculate the sample mean. The question that could be asked might be: 'Could the observations giving rise to this mean reasonably belong to that particular population?' We know that $\mu \pm 2\sigma$ will include 95% of the means of many samples of size n. If the observed mean lies outside this range, then it is said to differ from the true mean at the 5% level of significance. The level of significance was chosen arbitrarily to be 5%. It may be 10% ($\mu \pm 1.6\sigma$) or 1% ($\mu \pm 2.6\sigma$), or any percentage.

## The t-test

In the previous section the concept of hypothesis testing was explained in the unlikely situation wherein the population mean and variance are known. In practice we do not know the population variance and it has to be estimated from the data set using the formula:

$$s^2 = \{\Sigma X^2 - (\Sigma X)^2 / n\}/(n-1), \text{ or } \Sigma x^2 /(n-1), \text{ where } x = X - \overline{X}$$

The standard error of the mean would then be: se $SE = \sqrt{\Sigma x^2/(n-1)n}$. The difference between two means divided by the standard error of the means is called Student's t or simply t. It follows a distribution of the expected values of ratios such as the above originally worked out by V.S. Gosset, a statistician.

The difference between two means ($\overline{X}_1 - \overline{X}_2$) can be tested using Students' t:

$$t_{(n-1)} = (\overline{X}_1 - \overline{X}_2)/ \sqrt{(s_1^2/n_1 + s_2^2/n_2)}$$

Use of the t test to compare two samples of equal size is explained in the example given in Box II.20.

The t value at 13($n$–1) degrees of freedom as given in Annex 2, is 2.160 at $P \leq$ 0.05 (5% level of probability). We reject the hypothesis that there is no difference between the birth weight of male and female calves because the estimated t value (2.25) is greater than that given in Annex 2. Otherwise we state that the birth weights of the male and female calves differ significantly at the 5% level. The t test is not most commonly used for the comparison of two samples as shown above.

**Box II.20**   Example of doing a t – test

| The birth weights (kg) of 14 male and female calves are given below. Find out whether there is significant difference between the samples. |||||||||||||||
|---|---|---|---|---|---|---|---|---|---|---|---|---|---|---|

| Male | 33 | 42 | 28 | 34 | 30 | 36 | 31 | 32 | 33 | 30 | 31 | 23 | 29 | 28 |
|---|---|---|---|---|---|---|---|---|---|---|---|---|---|---|
| Female | 28 | 31 | 30 | 27 | 24 | 30 | 31 | 32 | 30 | 27 | 25 | 28 | 29 | 26 |

|  | Male | Female |
|---|---|---|
| Mean birth weight (kg) | 31.42 | 28.43 |
| $\Sigma X$ | 440.00 | 398.00 |
| $\Sigma X^2$ | 14,078.00 | 11,390.00 |
| $(\Sigma X)^2/n$ | 13,828.60 | 11,314.60 |
| $\Sigma X^2-(\Sigma X)^2/n$ | 249.40 | 75.40 |
| $\{\Sigma X^2-(\Sigma X)^2/n\}/(n-1)=s^2$ | 19.18 | 5.38 |

$$t = (\text{mean}_1 - \text{mean}_2)/\sqrt{s_1^2/n_1 + s_2^2/n_2} \quad (31.42 - 28.43) / \sqrt{(19.18/14 + 5.38/14} = 3/1.33$$
$$= \textbf{2.25}$$

**Box II.21**   Example of hypothesis testing when sample sizes differ

| The birth weight of 14 male and 10 female calves are given below. Find out whether there is significant difference between the samples. |||||||||||||||
|---|---|---|---|---|---|---|---|---|---|---|---|---|---|---|

| Male | 33 | 42 | 28 | 34 | 30 | 36 | 31 | 32 | 33 | 30 | 31 | 23 | 29 | 28 |
|---|---|---|---|---|---|---|---|---|---|---|---|---|---|---|
| Female | 28 | 31 | 30 | 27 | 24 | 30 | 31 | 32 | 30 | 27 |  |  |  |  |

|  | Male | Female |
|---|---|---|
| Mean birth weight (kg) | 31.42 | 29.00 |
| $\Sigma X$ | 440.00 | 290.00 |
| $\Sigma X^2$ | 14,078.00 | 8664.00 |
| $(\Sigma X)^2/n$ | 13,828.50 | 84.10 |
| $\Sigma X^2-(\Sigma X)^2/n$ | 249.40 | 54.00 |
| $\{\Sigma X^2-(\Sigma X)^2/n\}/(n-1)=s^2$ | 18.18 | 6.00 |

$$t = (\text{mean}_1 - \text{mean}_2)/\sqrt{s_1^2/n_1 + s_2^2/n_2} \quad (31.42 - 29.00)/\sqrt{(19.18/14 + 6.0/1.0} = 2.42 / 1.40$$
$$= \textbf{1.73}$$

The t value at 22($n_1+n_2-2$) degrees of freedom as given in Annex 2, is 2.086 at P $\leq 0.05$ (5 per cent level of probability). Therefore we accept the hypothesis that there is no difference between the birth weight of male and female calves. Otherwise we state that the birth weights of the male and female calves do not differ significantly at the 5% level.

# Analysis of variance (ANOVA)

Analysis of variance is used to look for differences among several groups of treatments. The procedure is explained using the following example (see Box II.22). The notations used in Box II.22 are:

| | |
|---|---|
| $n_i$ | Number of observations in each group is given separately in the appropriate columns |
| $\Sigma X_i$ | Sum of observations in each group is calculated and given against the respective groups |
| $\Sigma X_i^2$ | Square of each observation summed for the groups |
| $(\Sigma X_i)^2$ | Sum of the group $\Sigma(X_i)$ is squared and given in the respective columns |
| $(\Sigma X_i)^2 / n_i$ | Sum of the observations in each group $\Sigma(X_i)$ is squared and divided by the respective number of observations |

**Box II.22** Analysis variance to test differences in birth weight of calves belonging to four genetic groups

| Explanation | Notation | 50% BS | 62.5% BS | 75% BS | BS | Total |
|---|---|---|---|---|---|---|
| | | 32 | 33 | 33 | 34 | |
| | | 27 | 32 | 34 | 42 | |
| | | 25 | 28 | 31 | 38 | |
| | | 18 | 32 | 25 | 29 | |
| | | 32 | 30 | 26 | 28 | |
| | | 23 | 29 | 28 | 36 | |
| | | 25 | 30 | 22 | 44 | |
| | | 28 | | 30 | 43 | |
| | | 28 | | 33 | 30 | |
| | | | 21 | 39 | | |
| | | | 29 | 38 | | |
| | | | 27 | 40 | | |
| | | | | 41 | | |
| Number of observations | $n_i$ | 9 | 7 | 12 | 14 | N = 42 |
| Sum of observations | $\Sigma X_i$ | 238 | 214 | 339 | 513 | 1304 |
| Sum of each squared observation | $\Sigma X_i^2$ | 6448 | 6562 | 9775 | 19177 | 41962 |
| Squared sum of observations | $(\Sigma X_i)^2$ | 56644 | 45796 | 114921 | 263169 | |
| Squared sum of observations /number of observations | $(\Sigma X_i)^2/n$ | 6294 | 6542 | 9577 | 18798 | 41211 |

Contd.

| **Box II.22** Contd. | | | |
|---|---|---|---|
| Overall sum of squared observations/ overall number of observations | $(\Sigma\Sigma X_i)^2/N$ | | $1304^2/42 = 40486.1$ |

**Analysis of variance table (ANOVA)**

| Source | Df | Sum of squares | Mean square | F |
|---|---|---|---|---|
| Between breeds | 4–1 = 3 | $\Sigma[(\Sigma X_i)^2/n_i]-(\Sigma\Sigma X_i)^2/N$ <br> $41,211 - 40,486.1 = 724.9$ | $724.9 / 3 = 241.6$ | $241.6 / 19.76$ <br> $= 12.23$ |
| Within breeds | 42–4 = 38 | $\Sigma\Sigma X_i^2-\Sigma[(\Sigma X_i)^2/n_i]$ <br> $41,962 - 41,211 = 751$ | $751 / 38 = 19.76$ | |

F value from Annex 4 for $F_{38}^{3}$=4.31,2.84 (1% and 5% levels of significance respectively)
We observe that there are significant differences among the genetic groups.

Variance analysis is one of the most widely used statistical techniques in animal breeding. It is applied to situations wherein we look for differences among several treatments or groups of observations. The ratio of the mean square among groups (breeds) and the mean square within groups (breeds) is the F-statistic that can be compared to F values for different levels of significance tabulated in Annexes 4a and 4b with the degrees of freedom for the greater mean square horizontally and the lesser mean square vertically. If the value of the calculated F–statistic exceeds the values in Annex 4 to the corresponding degrees of freedom, then we say that the groups differ significantly at that level of significance.

From the expectations of the calculated mean squares it is possible to split the variance into components of variance, which is an often-used technique in quantitative genetics.

In the foregoing section making the reader follow the steps whereby one can do simple variance analysis was attempted. For more detailed and complicated calculations readers may refer to standard statistical textbooks.

## The chi square test

The test of significance based on the chi square distribution is used when discontinuous data are analysed. With the chi square test the observed frequency is compared to the expected. In a field performance recording programme, e.g., cows are milk recorded monthly with an accuracy of one decimal by milk recorders. We expect that in a normal unbiased situation, the numbers after the decimal point should be equally distributed between zero and nine. Now the result from a particular milk recorder who did 100 recordings in a month is analysed using the chi square test in Box II.23

**Box II.23** Chi square test to assess performance of milk recorders

| Frequency | 0.0 | 0.1 | 0.2 | 0.3 | 0.4 | 0.5 | 0.6 | 0.7 | 0.8 | 0.9 |
|---|---|---|---|---|---|---|---|---|---|---|
| Expected | 10 | 10 | 10 | 10 | 10 | 10 | 10 | 10 | 10 | 10 |
| Observed | 35 | 3 | 6 | 5 | 2 | 5 | 4 | 5 | 3 | 2 |
| (Observed– Expected)$^2$ | 625 | 49 | 16 | 25 | 64 | 625 | 36 | 25 | 49 | 64 |

Chi square value $(\chi^2) = \Sigma(\text{observed} - \text{expected})^2/\text{expected}$

$\chi^2 = (35 - 10)^2/10 + \dots\dots + (2 - 10)^2/10 = 157.8$

If all the observed values were exactly 10, $\chi^2$ will be 0. Therefore, the higher the value of $\chi^2$, the lower the probability that the result is by chance. The $\chi^2$-value computed in the above example is 157.8 and can be checked with the figures given in Annex 3 for 9 degrees of freedom. We can observe that the computed value of chi square is higher than any of the values shown in Annex 3 for 9 degrees of freedom. In our example this means the recordings are less reliable.

# Regression and correlation

## Regression

So far we have studied one variable at a time except in the case of chi square test wherein the association between two variables was studied in the form of counts. In this chapter the dependence of one variable, Y, on another variable, X, is studied. Two variables are often studied in the hope to attribute some underlying relationship between them. In reality we may come across two variables, one of which is dependent on the other. One of the most effective ways of understanding the relation between two variables is a scatter diagram. The example given in Box II. 24 will help in understanding this relationship.

**Box II.24**   Peak daily yield (kg) and fat % in Sunandini cows

| Fat % | 3.1 | 3.4 | 3.3 | 2.5 | 2.6 | 3.5 | 3.0 | 2.6 | 3.2 | 2.3 | 2.7 |
|---|---|---|---|---|---|---|---|---|---|---|---|
| Peak yield (kg) | 15 | 17 | 18 | 10 | 11 | 20 | 16 | 12 | 14 | 10 | 12 |

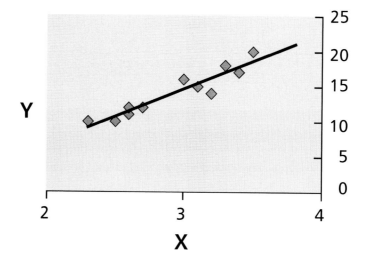

It is only intended in this chapter to introduce the principle behind regression analysis to the readers. Readers may refer to any standard textbook for a detailed understanding of regression and correlation. There are 11 points on the diagram, one for each individual. When a straight line is fitted through the centre of the scatter we observe that all the data are close to the straight line. This line is called the regression line. The regression line reveals the functional relationship between X and Y. The equation for the regression line is:

$$\hat{Y} = \overline{Y} + ß\,(X - \overline{X})$$

The symbol ß denotes the regression coefficient and is computed as:

$$ß = [\Sigma XY - (\Sigma X \cdot \Sigma Y/n)]/[\Sigma X^2 - (\Sigma X)^2/n]$$

The regression coefficient gives the change in estimating Y for a unit change in X and so is employed to estimate the Y from a given X.

In our example given in Box II.23, the regression coefficient is calculated as:

$$\Sigma XY = 466.7;\ \Sigma X\ \Sigma Y/n = 453.7;\ \Sigma X^2 = 95.9;\ (\Sigma X)^2/n = 94.3$$
$$(466.7 - 453.7)/(95.9 - 94.3) = 13/1.6 = 8.13$$

It is possible to estimate the value of Y for any given value of X by using the formula

$$\hat{Y} = \overline{Y} + ß(X - \overline{X}),\ \text{where } \hat{Y} \text{ is the estimate for Y}$$

## Correlation

In an ANOVA, we partition the variance of a trait or variable according to effects and we test for significant differences between levels. In the example in Box III.24, the effects consisted of different levels of Brown Swiss inheritance. (Alternatively, we could have estimated a regression on the level of Brown Swiss inheritance.) The degree of association or correlation between two variables is measured most commonly as the correlation coefficient r. The formula to calculate the correlation coefficient is:

$$R = \Sigma xy/\sqrt{\Sigma x \cdot \Sigma y}$$

where $\Sigma xy = \Sigma(X \cdot Y) - (\Sigma X \cdot \Sigma Y/n)$
$\Sigma x = \Sigma X^2 - (\Sigma X)^2/n$
$\Sigma y = \Sigma Y^2 - (\Sigma Y)^2/n$

The correlation coefficient can be either negative or positive. If positive, this means that the two variables (say traits) vary in the same direction, i.e. if one increases, the other also increases. When the correlation coefficient is negative, this means that when one variable increases the other decreases. Let us examine the example given in Box II.25.

In the formula for coefficient of correlation the denominator is always positive and the numerator always determines the sign (positive or negative). The coefficient of correlation always lies between –1 and + 1. A correlation coefficient of

**Box II.25**   Determination of correlation coefficient

In a herd of cows the first lactation peak daily yield of milk (kg) and the fat % are recorded and given below. Find out whether there is any correlation between the peak daily yield and the fat % in the herd.

In the following example, the correlation between milk yield and fat % is calculated.

| X | 13 | 12 | 18 | 21 | 16 | 20 | 12 | 10 | 16 | 12 | 19 | 17 | 14 | 17 | 14 | 19 | 22 | 23 |
|---|----|----|----|----|----|----|----|----|----|----|----|----|----|----|----|----|----|----|
| Y | 4.2 | 4.8 | 3.9 | 3.8 | 4 | 4.1 | 4.5 | 4.8 | 4 | 4.4 | 3.7 | 3.9 | 4 | 4.1 | 4.9 | 4 | 3.3 | 3.2 |
| X*Y | 55 | 58 | 70 | 80 | 64 | 82 | 54 | 48 | 64 | 53 | 70 | 66 | 56 | 70 | 69 | 76 | 73 | 74 |

|  | X (Daily yield kg) | Y (Fat %) | X·Y |
|---|---|---|---|
| $n$ | 18 | 18 | |
| Sum of observations | 295 | 73.6 | 1180.1 |
| Sum of square observations | 5083 | 304.64 | |
| $(Sum)^2$ | $295^2 = 87025$ | $73.6^2 = 5417$ | |
| $(Sum)^2/n$ | 4834.7 | 300.9 | |
| $\sum X \cdot \sum Y/n$ | $295 \cdot 73.6 / 18 = 1206.22$ | | |
| $\sum X^2 - (\sum X)^2/n = \sum x$ | $5083 - 4834.7 = 248.3$ | | |
| $\sum Y^2 - (\sum Y)^2/n = \sum y$ | $304.64 - 300.9 = 3.74$ | | |
| $\sum X \cdot \sum Y - \sum X \cdot \sum Y/n = \sum xy$ | $1180.1 \cdot 1206.2 = -26.1$ | | |
| Coefficient of correlation $(r) = \sum xy / \sum x \cdot \sum y$ | $-26.1 / (248.3 \cdot 3.74) = -0.856$ | | |

zero means that there is no linear relation between the two variables, either positive or negative. If the correlation coefficient is 0.50, this means that 50% of the variance in one variable is associated with the variance of the other and the remaining is not.

# References

Adams, R.H. 1996. *Livestock Income, Male/Female Animals, and Inequality in Rural Pakistan.* Food Consumption and Nutrition Division. International Food Policy Research Institute. Washington OC, USA.

Ahuja, V. 2004. *Livestock and Livelihoods: Challenges and Opportunities for Asia in the Emerging Market Environment.* National Dairy Development Board, Anand, India. Pro-Poor Livestock Policy Facility (South Asia Hub),New Delhi, India FAO, Rome, Italy.

Ahuja, V., George, P.S., Ray, S. Mc Connell, K.E., Kurup, M.P.G., Gandhi, V., Dieninger, D.U. and de Haan, C. 2000. *Agricultural Services and the Poor – Case of Livestock Health and Breeding Service in India.* Indian Institute of Management, Ahmedabad, The World Bank and Swiss Agency for Development and Cooperation.

AHS Series-9, 2004. Basic Animal Husbandry Statistics 2004. Government of India. Ministry of Agriculture. Department of Animal Husbandry & Dairying, New Delhi, India.

Alexander, G.I. and Tierney, M.L. (1990). Selection methods used in ABS breed of tropical dairy cattle. Proc. of the 4th World congress on genetics applied to livestock production, Edinburgh 23-27 July 1990. XIV: 263 - 266. Conference paper.

Alstrom, S. 1977. The dairy crossbreeding programme in India. FAO report (unpublished).

Annual Report. 2003. Department of Animal Husbandry and Dairying Government of India, New Delhi, India.

Bhosrekar, M.R. 1988. Problems involved in freezing crossbred bull semen. *A.I. News 3: 20 – 24.*

Bhosrekar, M.R. 2003. Managing the breeding bull. *Compendium of the National Seminar on Frozen Semen Technology.* Dhoni, Kerala, India, pp. 56 – 63.

Bichard, M. 1999. Summary of the workshop outcome. ICAR Technical, series no. 3. *Workshop on Developing Breeding Strategies for Low Input Animal Production Environments.* Bella, Italy, pp. 5 – 12.

Burman, A.R. 2003. Maximising productivity of bulls and good laboratory practices. *Compendium of National Seminar on Frozen Semen Technology.* Dhoni, Kerala, India, pp. 73 – 79.

Chacko, C.T. 1980. The genetic consequences of different selection policies for crossbred bulls for use in AI in Kerala (India). Diss., Institute of Genetics, Univ. of Edinburgh, UK.

Chacko, C.T. 1985. Consequences of different selection application programmes in genetic improvement of crossbred cattle in Kerala. *Indian Vet. J* 62:591 – 600.

Chacko, C.T. 1989. A sire evaluation system for crossbred bulls in India. *Proc. 23rd Dairy Industry Congress,* Hyderabad, India.

Chacko,C.T. 1992. Progeny testing of crossbred bulls in tropics—The Kerala programme. A case study. *Compendium of National Seminar of Progeny Testing of Bulls in Tropics.* Thiruvanathupuram, Kerala, India.

Chacko, C.T. 1999. An approach to bovine breeding programme in India. Workshop on breed conservation and development. Lucknow, India, November 10,1999.

Chacko, C.T. 2003. Bull selection for genetic improvement: a practical approach for Indian situation. *Compendium of National Seminar on Frozen Semen Technology.* Dhoni, Kerala, India.

Chacko, C.T., Bachmann, F. and Kropf, W. 1985. Results of dairy improvement programmes under field conditions in Kerala. *36th annual meeting of the EAAP.* Kallithea, Greece.

Chacko, C.T., Schneider, F. and Ramachandran, U. 1988. Progeny testing in a large scale crossbreeding programme in Kerala: scope and limitations. *39th annual meeting of the EAAP.* Helsinki, Finland.

Coulter, G.H., Ville, T.R. and Foote, R.H. 1976. *J. Anim. Sci.* 43: 9.

Cunningham, E.P. and Syrstad, O. 1987. Crossbreeding *Bos indicus* and *Bos taurus* for milk production in tropics. FAO Animal Production and Health, Paper 68.

Delgado, C., Rosegrant, M., and Meijer, S. 2001. Livestock to 2020: the revolution continues. Paper presented at the annual meetings of the International Agricultural Trade Research Consortium (IATRC). Auckland, New Zealand.

Delgado, C., Rosegrant, M., Steinfeld, H., Ehui, S., and Courbois, C. 1999. Livestock to 2020. The next food revolution. IFPRI, FAO, ILRI. Food, Agriculture and Environment Paper 28. Washington DC, USA.

Devanand, C. P., Krishnan, R., Biju, K., Aravaind, T. P. and Rejesh, V. R. 2003. Quality assurance for bovine frozen semen. *Compendium of National Seminar on Frozen Semen Technology,* Dhoni, Kerala, India, pp. 174 – 194.

Falconer, D. S. 1989. *Introduction to Quantitative Genetics.* Longman Scientific and Technical Publ., Harlow, Essex. (3rd ed.).

FAO Statistics Database. 2005. FAOSTAT Agriculture Data (http://www.fao.org/Last accessed April 2005, FAO, Rome, Italy).

George, P.S. and Nair K.N. 1990. Livestock economy of Kerala. Centre for Development Studies, Trivandrum, Kerala, India.

George, P.S., Nair,K.N., Strebel, B., Unnithan, N.R., and Wälty, S. 1989. Policy options for cattle development in Kerala. Centre for Development Studies. Trivandrum, Kerala, India.

Government of Kerala. 2003. Economic review by the state planning board, Thiruvanathupuram, Kerala, India.

Gurnani, M. 1981. Estimates of genetic and environmental trends in milk production of dairy cattle. Ph D thesis, Punjab University, Chandigarh, Punjab, India.

Haard, M. (1997). Semen processing: general aspects, new diluents, cooling rates. Proc. of the 9th European AI meeting, 29 - 33.

Hafs, H.D. (1972). Management of bulls to maximise sperm output. Proc. of the 4th Technical conference on artificial insemination and reproduction, 44. NAAB.

Hall, D. 2004. Implications for policy making. Summary report. *FAO – LEAD Workshop. 2004.* Bangkok. Thailand.

Hill, W. G. 1974. Prediction and evaluation of response to selection with overlapping generations. *Anim. Prod.* 18: 117.

Hinsch, E., Hinsch, K.D., Boehm, J.G., Schill, W.B. and Schloesser, M.F. (1997). Functional parameters and fertilization success of bovine semen cryopreserved in egg yolk free and egg yolk containing extenders. Journal of Reproduction in domestic animals. 32: 149. Giessen. Germany.

Hodges, J. (1986). Animal genetic resources in the developing world; goals, strategies, management and current status. Proceedings of the 3rd World congress on genetics applied to livestock production. July 16 to 22, 1986. Lincoln, Nebraska, USA. XII 474 - 485.

Hodges, J. 1990. Conservation of animal genetic resources in developing countries. Genetic conservation of domestic livestock. *C.A.B. International,* UK, pp. 128 – 145.

IFAD. 2004. Livestock services and the poor. A global initiative. Collecting, coordinating and sharing experiences. DANIDA, World Bank, IFAD, Rome, Italy.

ICAR (International Council for Animal Recording) 1995. International Agreement of Recording Practices. ICAR, Rome, Italy.

ICAR. 1999. Technical series no. 3. *Workshop on Developing Breeding Strategies for Low Input Animal Production Environments.* Bella, Italy, 563 pp.

ICUN, 1980. *World Conservation Strategy: Living Resource Conservation for Sustainable Development.* IUCN, Gland, Switzerland.

Intercooperation. 2000. Capitalisation of experiences in livestock production and dairying (LPD) in India. *Technical Report 15.* Intercooperation. Bern, Switzerland.

Jose, T.K. and Chacko, C.T. 1992. Progeny testing programme in Kerala. *National Seminar on Progeny Testing of Bulls in Tropics,* Kerala Livestock Development Board, Trivandrum, India.

Kalev, G and D. Zagorzki, 1968. Deep freezing of bull semen. *Anim Breeding Abstracts.* 40, no 4484.

Krishnamurthy, S. 1993. The indigenous breeds of cattle and buffaloes in India: Present status and future outlook. Internal working paper. SDC, New Delhi. India.

Krook, L., Lutwak L., McEntee, K., Brann, K. and Roberts, S. (1972). The effect of high dietary calcium in bulls. Proceedings of the 4th Technical conference on artificial insemination and reproduction 65 - 67. NAAB.

Kropf, W. and Chacko, C.T. 1992. Merits of using exotic cattle breeds for performance improvement in the tropics. *Compendium of 43rd Annual Meeting of the EAAPO.* Madrid, Spain. Session G+C, 123.

Kunzi, N. and Kropf, W. (1986). Genetic improvement for milk and meat production in tropics. Proceedings of the 3rd World congress on genetics applied to livestock production. July 16 to 22, 1986. Lincoln, Nebraska, USA. IV: 254 - 262.

Kurup M.P.G. 1995. Livestock sector in India, an analysis and overview. Internal working paper. SDC, New Delhi, India.

Kurup, M.P.G. 1998a. CAPEX: Context – India, Chronological tables 1868 – 1998 and explanatory notes on identified events/issues. Input paper for 2nd CAPEX workshop, Bangalore-Attibele, Karnataka. India.

Kurup, M.P.G. 1998b. National project for cattle and buffalo breeding. Doorstep delivery of breeding services and total coverage of breedable cows and buffaloes. *Project report. DAHD.* Department of Animal Husbandry and Dairying, Ministry of Agriculture, Government of India. New Delhi. India.

Kurup. M.P.G. 2003. *Livestock in Orissa. The Socio-Economic Perspective.* Manohar Publishers and Distributors, New Delhi, India.

Lamond, D.R. and Campbell, E.A. 1970. *Dairy Cattle Husbandry.* Angus and Robertson, Sydney, Australia, 126 pp.

Land, R.B. (1986). Genetic resource requirements under favourable production marketing systems; priorities and organisation. Proceedings of the 3rd World congress on applied livestock production. July 16 to 22, 1986. Lincoln, Nebraska, XII: 486 - 491.

Lehmann, R., Vishva, R., Ramesh, K.S., Nandita, R., Subrahmaniam, S., and Wälty, S. 1994. Bovine and dairy development in Andhra Pradesh. Indo-Swiss Project Andhra Pradesh, Hyderabad, India.

LEAD (Livestock, Environment and Development Initaitive). 2004. http://www.lead.virtualcentre.org. FAO. Rome. Italy.

Lush, J. L. 1950. Genetics and animal breeding. In: *Genetics in the Twentieth Century*, L.C. Dun (ed.). Macmillian Co., New York, NY.

Maijala, K. 1982. Preliminary report of the working party on animal genetic resources in Europe. In: *Conservation of Animal Genetic Resources, Session 1. Commission on Animal Genetics*, EAAP, G.1.2. Leningrad.

Mason I. L. 1974. Maintaining crossbred populations of dairy cattle in the tropics. *World Anim. Rev.* FAO 11: 36 – 43.

Mathew, A. (1984). Principles and practices of deep freezing bull semen. Training Centre, KLD & MM Board, Mattupatti, Kerala, India.

Menzi, M. (1993). Farmer's participation in cattle development in the rural areas. Compendium of the National Seminar on Frozen semen technology. Mattupatti, Kerala, India, 44 - 46.

Morrenhof, J., Ahuja, V., Tripathy, A. (2004). Livestock services and the poor. Paper proceedings and presentations of the international workshop, held in Bhubaneswar, India, October 28 - 29, 2002. Swiss Agency for Development and Cooperation, Bern, Switzerland and FAO, Rome, Italy.

Nair, P. N. R. 1973. Evolutionary crossbreeding as a basis for cattle development in Kerala state (India). Thesis, University of Zurich, Faculty of Veterinary Medicine, Zuxich.

NDDB (National Dairy Development Board). 2002. Report on the breeding policies and programmes in India. NDDB, Anand, Gujarat, India.

NAS (National Account Statistics). 2000 Central Statistical Organisation. Government of India, New Delhi, India.

NLP (National Livestock Policy). 1996. Report of the Steering Group. DAHD (Department of Animal Husbandry and Dairying). Ministry of Agriculture. Government of India. New Delhi. India.

National Sample Survey Organisation (NSSO). 1992. *Land and livestock holding survey: NSS 48th Round*. NSSO Report 408. http://www.nic.in/stat/stat_act_t13.htm. New Delhi, India.

Nehring, H. and Rothe, L. 2003. Insemination of cryopreserved bull semen portions with reduced sperm numbers after dilution with two egg yolk free extenders. *Proceedings of 15 European A.I. Vets meeting*. Budapest (Hungary). pp. 14 – 23.

Nitter, G. 1999. Developing crossbreeding structures for extensive grazing systems, utilising only indigenous animal genetic resources. Working document. *Workshop on Developing Breeding Strategies for Lower Input Animal Production Environments*, FAO, ICAR, SDC and GTZ. Bella, Italy.

Rajamannan, A.H.J., Gram, E. F. and Schmehal, M.K.L. 1971. Effect of holding time on bovine semen. *Artificial Insemination Digest.* 19(4): 6.

Reklewski, Z., Jasiorowski, H. and Stolzman, M. 1984. Investigations of performance of ten groups of Friesian cattle in an international comparison. *Tierzuchter* 36: (7): 288 – 291

Rendel, J.M. and Robertson, Alan. 1950. Estimation of genetic gain in milk yield by selection in a closed herd of dairy cattle. *J. Genetics* 50:1.

Robertson, A. 1980. *Notes on Animal Breeding*. Institute of Animal Genetics. Edinburgh, UK.

Schneider, F. 1999. Livestock and the Environment. *InfoAgrar News*, No. 5, Zollikofen, Switzerland.

Schneider, F., Goe M.R., Chacko, C.T., Wieser, M., and Mulder, H. 1999. Field performance recording in smallholder dairy production systems in India. Paper. 50th Annual EAAP Meeting, 22 – 26 August, Zürich, Switzerland.

SDC/IC (Swiss Agency for Development and Cooperation and Intercooperation). 1995a. Developing a partnership. Planning pre-phases as an instrument in project planning. Internal working paper. New Delhi, India.

SDC/IC 1995b. Sector strategy for programme development in livestock production and dairying in India. Internal working paper. New Delhi, India.

Singh, S.K. and Nagarsankar,R. 2000. Estimation of genetic gain in milk yield by selection in Sahiwal herds in India. *Indian J. Anim. Sci.* 70:286 – 288.

Smith, C. 1984. Rate of genetic change in farm livestock. *Res. and Devel. in Agric.* I: 79 – 85.

Snedecor, G.W. and Cochran, W.G. 1967. *Statistical Methods.* Oxford and IBH Publ. Co., New Delhi (6th ed.).

Steinfeld, H., Blackburn, H., and de Haan, K. 1997. *Livestock Environment Interactions. Issues and Option.* Report on a study sponsored by EU, WB, DK, F, D, NL, UK, USA and FAO. Rome, Italy.

Sudheer, S. 2000. Relationship between testicular size and seminal attributes in crossbred bulls. *Indian J. Anim. Reprod.* 34(2): 159 – 60.

Sudheer, S., Xavier, C .J., Kizhakudan, J. and Varghese, K. 2001. Breeding life and culling of crossbred AI bulls in the tropics. *Indian Vet. Med. J.* 25: 125 – 126.

Taneja, V.K. 1999. Cattle breeding programmes in India. ICAR Technical series no. 3. *Workshop on Developing Breeding Strategies for Low Input Animal Production Environments.* FAO, ICAR, SDC, GTZ, Bella, Italy; pp 445 – 454.

Taneja, V. K. and Chawla, D. S. 1978. Comparative study of economic traits of Brown Swiss x zebu crossbreds. *Indian J. Dairy Sci.* 31(2): 188–190.

Tewolde, A. 1999. Breeding strategies for low input animal production systems: a case study from Central America and Mexico. ICAR Technical series no. 3. *Workshop on Developing Breeding Strategies for Low Input Animal Production Environments.* FAO, ICAR, SDC, GTZ, Bella, Italy; pp. 387 – 394.

Trivedi, K.R. 2000. Formulation of strategies for genetic improvement of cattle and buffaloes. *Report National Dairy Development Board,* Anand, Gujarat, India.

Vaccaro, L.P. (1979). The performance of dairy cattle breeds in tropical Latin America and programmes for their improvement. Dairy cattle breeding in humid tropics. Hissar, India, 121 - 136. Working paper FAO Rome, Italy and GOI New Delhi, India.

Van der Werf, J. 1999. Livestock straight breeding system structures for sustainable intensification of extensive grazing systems. ICAR Technical series no. 3. *Workshop on Developing Breeding Strategies for Low Input Animal Production Environments,* FAO, ICAR, SDC, GTZ, Bella, Italy, pp. 105 – 177.

Wälty, S. 1999. Cows, buffaloes and the rural poor in India. *J Entwicklungspolitik.* Wien, Austria.

Weller, J. 1999. Economic evaluation of straight- and cross-breeding programs. *Working document. Workshop on Developing Breeding Strategies for Low Input Animal Production Environments.* FAO, ICAR, SDC, GTZ, Bella. Italy.

World Bank. 1996. India livestock sector review: enhancing growth and development. Report No. 14522-IN. *Agriculture and water operations division, country department II, South Asia Region.* Washington DC USA.

# Annexes

**Annex 1.** Standard normal distribution

| z | λ | z | λ | z | λ | z | λ | z | λ |
|------|-------|------|-------|------|-------|------|-------|------|-------|
| .01 | .5040 | .51 | .6950 | 1.01 | .8438 | 1.51 | .9345 | 2.01 | .9778 |
| .02 | .5080 | .52 | .6985 | 1.02 | .8461 | 1.52 | .9357 | 2.02 | .9783 |
| .03 | .5120 | .53 | .7019 | 1.03 | .8485 | 1.53 | .9370 | 2.03 | .9788 |
| .04 | .5160 | .54 | .7054 | 1.04 | .8508 | 1.54 | .9382 | 2.04 | .9793 |
| .05 | .5199 | .55 | .7088 | 1.05 | .8531 | 1.55 | .9394 | 2.05 | .9798 |
| .06 | .5239 | .56 | .7123 | 1.06 | .8554 | 1.56 | .9406 | 2.06 | .9803 |
| .07 | .5279 | .57 | .7157 | 1.07 | .8577 | 1.57 | .9418 | 2.07 | .9808 |
| .08 | .5319 | .58 | .7190 | 1.08 | .8599 | 1.58 | .9429 | 2.08 | .9812 |
| .09 | .5359 | .59 | .7224 | 1.09 | .8621 | 1.59 | .9441 | 2.09 | .9817 |
| .10 | .5398 | .60 | .7257 | 1.10 | .8643 | 1.60 | .9452 | 2.10 | .9821 |
| .11 | .5438 | .61 | .7291 | 1.11 | .8665 | 1.61 | .9463 | 2.11 | .9826 |
| .12 | .5478 | .62 | .7324 | 1.12 | .8686 | 1.62 | .9474 | 2.12 | .9830 |
| .13 | .5517 | .63 | .7357 | 1.13 | .8708 | 1.63 | .9484 | 2.13 | .9834 |
| .14 | .5557 | .64 | .7389 | 1.14 | .8729 | 1.64 | .9495 | 2.14 | .9838 |
| .15 | .5596 | .65 | .7422 | 1.15 | .8749 | 1.65 | .9505 | 2.15 | .9842 |
| .16 | .5636 | .66 | .7454 | 1.16 | .8770 | 1.66 | .9515 | 2.16 | .9846 |
| .17 | .5675 | .67 | .7486 | 1.17 | .8790 | 1.67 | .9525 | 2.17 | .9850 |
| .18 | .5714 | .68 | .7517 | 1.18 | .8810 | 1.68 | .9535 | 2.18 | .9854 |
| .19 | .5753 | .69 | .7549 | 1.19 | .8830 | 1.69 | .9545 | 2.19 | .9857 |
| .20 | .5793 | .70 | .7580 | 1.20 | .8849 | 1.70 | .9554 | 2.20 | .9861 |
| .21 | .5832 | .71 | .7611 | 1.21 | .8869 | 1.71 | .9564 | 2.21 | .9864 |
| .22 | .5871 | .72 | .7642 | 1.22 | .8888 | 1.72 | .9573 | 2.22 | .9868 |
| .23 | .5910 | .73 | .7673 | 1.23 | .8907 | 1.73 | .9582 | 2.23 | .9871 |
| .24 | .5948 | .74 | .7704 | 1.24 | .8925 | 1.74 | .9591 | 2.24 | .9875 |
| .25 | .5987 | .75 | .7734 | 1.25 | .8944 | 1.75 | .9599 | 2.25 | .9878 |
| .26 | .6026 | .76 | .7764 | 1.26 | .8962 | 1.76 | .9608 | 2.26 | .9881 |
| .27 | .6064 | .77 | .7794 | 1.27 | .8980 | 1.77 | .9616 | 2.27 | .9884 |
| .28 | .6103 | .78 | .7823 | 1.28 | .8997 | 1.78 | .9625 | 2.28 | .9887 |
| .29 | .6141 | .79 | .7852 | 1.29 | .9015 | 1.79 | .9633 | 2.29 | .9890 |
| .30 | .6179 | .80 | .7881 | 1.30 | .9032 | 1.80 | .9641 | 2.30 | .9893 |
| .31 | .6217 | .81 | .7910 | 1.31 | .9049 | 1.81 | .9649 | 2.31 | .9896 |
| .32 | .6255 | .82 | .7939 | 1.32 | .9066 | 1.82 | .9656 | 2.32 | .9898 |
| .33 | .6293 | .83 | .7967 | 1.33 | .9082 | 1.83 | .9664 | 2.33 | .9901 |
| .34 | .6331 | .84 | .7995 | 1.34 | .9099 | 1.84 | .9671 | 2.34 | .9904 |
| .35 | .6368 | .85 | .8023 | 1.35 | .9115 | 1.85 | .9678 | 2.35 | .9906 |

Contd.

**Annex I.** (Continued)

| z | λ | z | λ | z | λ | z | λ | z | λ |
|---|---|---|---|---|---|---|---|---|---|
| .36 | .6406 | .86 | .8051 | 1.36 | .9131 | 1.86 | .9686 | 2.36 | .9909 |
| .37 | .6443 | .87 | .8078 | 1.37 | .9147 | 1.87 | .9693 | 2.37 | .9911 |
| .38 | .6480 | .88 | .8106 | 1.38 | .9162 | 1.88 | .9699 | 2.38 | .9913 |
| .39 | .6517 | .89 | .8133 | 1.39 | .9177 | 1.89 | .9706 | 2.39 | .9916 |
| .40 | .6554 | .90 | .8159 | 1.40 | .9192 | 1.90 | .9713 | 2.40 | .9918 |
| .41 | .6591 | .91 | .8186 | 1.41 | .9207 | 1.91 | .9719 | 2.41 | .9920 |
| .42 | .6628 | .92 | .8212 | 1.42 | .9222 | 1.92 | .9726 | 2.42 | .9922 |
| .43 | .6664 | .93 | .8238 | 1.43 | .9236 | 1.93 | .9732 | 2.43 | .9925 |
| .44 | .6700 | .94 | .8264 | 1.44 | .9251 | 1.94 | .9738 | 2.44 | .9927 |
| .45 | .6736 | .95 | .8289 | 1.45 | .9265 | 1.95 | .9744 | 2.45 | .9929 |
| .46 | .6772 | .96 | .8315 | 1.46 | .9279 | 1.96 | .9750 | 2.46 | .9931 |
| .47 | .6808 | .97 | .8340 | 1.47 | .9292 | 1.97 | .9756 | 2.47 | .9932 |
| .48 | .6844 | .98 | .8365 | 1.48 | .9306 | 1.98 | .9761 | 2.48 | .9934 |
| .49 | .6879 | .99 | .8389 | 1.49 | .9319 | 1.99 | .9767 | 2.49 | .9936 |
| .50 | .6915 | 1.00 | .8413 | 1.50 | .9332 | 2.00 | .9772 | 2.50 | .9938 |
| 2.51 | .9940 | 2.81 | .9975 | 3.11 | .9991 | 3.41 | .9997 | 3.71 | .9999 |
| 2.52 | .9941 | 2.82 | .9976 | 3.12 | .9991 | 3.42 | .9997 | 3.72 | .9999 |
| 2.53 | .9943 | 2.83 | .9977 | 3.13 | .9991 | 3.43 | .9997 | 3.73 | .9999 |
| 2.54 | .9945 | 2.84 | .9977 | 3.14 | .9992 | 3.44 | .9997 | 3.74 | .9999 |
| 2.55 | .9946 | 2.85 | .9978 | 3.15 | .9992 | 3.45 | .9997 | 3.75 | .9999 |
| 2.56 | .9948 | 2.86 | .9979 | 3.16 | .9992 | 3.46 | .9997 | 3.76 | .9999 |
| 2.57 | .9949 | 2.87 | .9979 | 3.17 | .9992 | 3.47 | .9997 | 3.77 | .9999 |
| 2.58 | .9951 | 2.88 | .9980 | 3.18 | .9993 | 3.48 | .9997 | 3.78 | .9999 |
| 2.59 | .9952 | 2.89 | .9981 | 3.19 | .9993 | 3.49 | .9998 | 3.79 | .9999 |
| 2.60 | .9953 | 2.90 | .9981 | 3.20 | .9993 | 3.50 | .9998 | 3.80 | .9999 |
| 2.61 | .9955 | 2.91 | .9982 | 3.21 | .9993 | 3.51 | .9998 | 3.81 | .9999 |
| 2.62 | .9956 | 2.92 | .9982 | 3.22 | .9994 | 3.52 | .9998 | 3.82 | .9999 |
| 2.63 | .9957 | 2.93 | .9983 | 3.23 | .9994 | 3.53 | .9998 | 3.83 | .9999 |
| 2.64 | .9959 | 2.94 | .9984 | 3.24 | .9994 | 3.54 | .9998 | 3.84 | .9999 |
| 2.65 | .9960 | 2.95 | .9984 | 3.25 | .9994 | 3.55 | .9998 | 3.85 | .9999 |
| 2.66 | .9961 | 2.96 | .9985 | 3.26 | .9994 | 3.56 | .9998 | 3.86 | .9999 |
| 2.67 | .9962 | 2.97 | .9985 | 3.27 | .9995 | 3.57 | .9998 | 3.87 | .9999 |
| 2.68 | .9963 | 2.98 | .9986 | 3.28 | .9995 | 3.58 | .9998 | 3.88 | .9999 |
| 2.69 | .9964 | 2.99 | .9986 | 3.29 | .9995 | 3.59 | .9998 | 3.89 | 1.0000 |
| 2.70 | .9965 | 3.00 | .9986 | 3.30 | .9995 | 3.60 | .9998 | 3.90 | 1.0000 |

Contd.

**Annex 1.** (Continued)

| z | λ | z | λ | z | λ | z | λ | z | λ |
|---|---|---|---|---|---|---|---|---|---|
| 2.71 | .9966 | 3.01 | .9987 | 3.31 | .9995 | 3.61 | .9998 | 3.91 | 1.0000 |
| 2.72 | .9967 | 3.02 | .9987 | 3.32 | .9996 | 3.62 | .9999 | 3.92 | 1.0000 |
| 2.73 | .9968 | 3.03 | .9988 | 3.33 | .9996 | 3.63 | .9999 | 3.93 | 1.0000 |
| 2.74 | .9969 | 3.04 | .9988 | 3.34 | .9996 | 3.64 | .9999 | 3.94 | 1.0000 |
| 2.75 | .9970 | 3.05 | .9989 | 3.35 | .9996 | 3.65 | .9999 | 3.95 | 1.0000 |
| 2.76 | .9971 | 3.06 | .9989 | 3.36 | .9996 | 3.66 | .9999 | 3.96 | 1.0000 |
| 2.77 | .9972 | 3.07 | .9989 | 3.37 | .9996 | 3.67 | .9999 | 3.97 | 1.0000 |
| 2.78 | .9973 | 3.08 | .9990 | 3.38 | .9996 | 3.68 | .9999 | 3.98 | 1.0000 |
| 2.79 | .9974 | 3.09 | .9990 | 3.39 | .9997 | 3.69 | .9999 | 3.99 | 1.0000 |
| 2.80 | .9974 | 3.10 | .9990 | 3.40 | .9997 | 3.70 | .9999 | | |

*(Reproduced from Snedecor and Cochran, 1967)*

**Annex 2.** Student's t. Values exceeded with probability P

| df | P=0.1 | 0.05 | 0.01 | df | P=0.1 | 0.05 | 0.01 |
|---|---|---|---|---|---|---|---|
| 1 | 6.314 | 12.706 | 63.657 | 18 | 1.734 | 2.101 | 2.878 |
| 2 | 2.920 | 4.303 | 9.925 | 19 | 1.729 | 2.093 | 2.861 |
| 3 | 2.353 | 3.182 | 5.841 | 20 | 1.725 | 2.086 | 2.845 |
| 4 | 2.132 | 2.776 | 4.604 | 21 | 1.721 | 2.080 | 2.831 |
| 5 | 2.015 | 2.571 | 4.032 | 22 | 1.717 | 2.074 | 2.819 |
| 6 | 1.943 | 2.447 | 3.707 | 23 | 1.714 | 2.069 | 2.807 |
| 7 | 1.895 | 2.365 | 3.499 | 24 | 1.711 | 2.064 | 2.797 |
| 8 | 1.860 | 2.306 | 3.355 | 25 | 1.708 | 2.060 | 2.787 |
| 9 | 1.833 | 2.262 | 3.250 | 26 | 1.706 | 2.056 | 2.779 |
| 10 | 1.812 | 2.228 | 3.169 | 27 | 1.703 | 2.052 | 2.771 |
| 11 | 1.796 | 2.201 | 3.106 | 28 | 1.701 | 2.048 | 2.763 |
| 12 | 1.782 | 2.179 | 3.055 | 29 | 1.699 | 2.045 | 2.756 |
| 13 | 1.771 | 2.160 | 3.012 | 30 | 1.697 | 2.042 | 2.750 |
| 14 | 1.761 | 2.145 | 2.977 | 40 | 1.684 | 2.021 | 2.704 |
| 15 | 1.753 | 2.131 | 2.947 | 60 | 1.671 | 2.000 | 2.660 |
| 16 | 1.746 | 2.120 | 2.921 | 120 | 1.658 | 1.980 | 2.617 |
| 17 | 1.740 | 2.110 | 2.898 | $\alpha$ | 1.645 | 1.960 | 2.576 |

*(Reproduced from Snedecor and Cochran, 1967)*

**Annex 3.** Cumulative distribution of chi square

| Degrees of freedom | Probability of a greater value | | | | |
|---|---|---|---|---|---|
| | .900 | .500 | .100 | .050 | .010 |
| 1 | .02 | .45 | 2.71 | .84 | 6.63 |
| 2 | .21 | 1.39 | 4.61 | 5.99 | 9.21 |
| 3 | .58 | 2.37 | 6.25 | 7.81 | 11.34 |
| 4 | 1.06 | 3.36 | 7.78 | 9.49 | 13.28 |
| 5 | 1.61 | 4.35 | 9.24 | 11.07 | 15.09 |
| 6 | 2.20 | 5.35 | 10.64 | 12.59 | 16.81 |
| 7 | 2.83 | 6.35 | 12.02 | 14.07 | 18.48 |
| 8 | 3.49 | 7.34 | 13.36 | 15.51 | 20.09 |
| 9 | 4.17 | 8.34 | 14.68 | 16.92 | 21.67 |
| 10 | 4.87 | 9.34 | 15.99 | 18.31 | 23.21 |
| 11 | 5.58 | 10.34 | 17.28 | 19.68 | 24.72 |
| 12 | 6.30 | 11.34 | 18.55 | 21.03 | 26.22 |
| 13 | 7.04 | 12.34 | 19.8 | 122.36 | 27.69 |
| 14 | 7.79 | 13.34 | 21.06 | 23.68 | 29.14 |
| 15 | 8.55 | 14.34 | 2.31 | 25.00 | 30.58 |
| 16 | 9.31 | 15.34 | 23.54 | 26.30 | 32.00 |
| 17 | 10.09 | 16.34 | 24.77 | 27.59 | 33.41 |
| 18 | 10.86 | 17.34 | 25.99 | 28.87 | 34.81 |
| 19 | 11.65 | 18.34 | 27.20 | 30.14 | 36.19 |
| 20 | 12.44 | 19.34 | 28.41 | 31.41 | 37.57 |
| 21 | 13.24 | 20.34 | 29.62 | 32.67 | 38.93 |
| 22 | 14.04 | 21.34 | 30.81 | 33.92 | 40.29 |
| 23 | 14.85 | 22.34 | 32.01 | 35.17 | 41.64 |
| 24 | 15.66 | 23.34 | 33.20 | 36.42 | 42.98 |
| 25 | 16.47 | 24.34 | 34.38 | 37.65 | 44.31 |
| 26 | 17.29 | 25.34 | 35.56 | 38.89 | 45.64 |
| 27 | 18.11 | 26.34 | 36.74 | 40.11 | 46.96 |
| 28 | 18.94 | 27.34 | 37.92 | 41.34 | 48.28 |
| 29 | 19.77 | 28.34 | 39.09 | 42.56 | 49.59 |
| 30 | 20.60 | 29.34 | 40.26 | 43.77 | 50.89 |
| 40 | 29.05 | 39.34 | 51.80 | 55.76 | 63.69 |
| 50 | 37.69 | 49.33 | 63.17 | 67.50 | 76.1 |
| 60 | 46.46 | 59.33 | 74.40 | 79.08 | 88.38 |
| 70 | 55.33 | 69.33 | 85.53 | 90.53 | 100.42 |
| 80 | 64.28 | 79.33 | 96.58 | 101.88 | 112.33 |
| 90 | 73.29 | 89.33 | 107.56 | 113.14 | 124.12 |
| 100 | 82.36 | 99.33 | 118.50 | 124.34 | 135.81 |

*(Reproduced from Snedecor and Cochran, 1967)*

**Annex 4a.** Variance ratio F. Probability = 0.05

| | 1 | 2 | 3 | 4 | 5 | 6 | 7 | 8 | 9 | 10 | 12 | 15 | 20 | 24 | 30 | 40 | 60 | 120 | ∞ |
|---|---|---|---|---|---|---|---|---|---|---|---|---|---|---|---|---|---|---|---|
| 1 | 161.4 | 199.5 | 215.7 | 224.6 | 230.2 | 234.0 | 263.8 | 238.9 | 240.5 | 241.9 | 243.9 | 245.9 | 248.0 | 249.1 | 250.1 | 251.1 | 252.2 | 253.3 | 254.3 |
| 2 | 18.51 | 19.00 | 19.16 | 19.25 | 19.30 | 19.33 | 19.35 | 19.37 | 19.38 | 19.40 | 19.41 | 19.43 | 19.45 | 19.45 | 19.46 | 19.47 | 19.48 | 19.49 | 19.50 |
| 3 | 10.13 | 9.55 | 9.28 | 9.12 | 9.01 | 8.94 | 8.89 | 8.85 | 8.81 | 8.79 | 8.74 | 8.70 | 8.66 | 8.64 | 8.62 | 8.59 | 8.57 | 8.55 | 8.53 |
| 4 | 7.71 | 6.94 | 6.59 | 6.39 | 6.26 | 6.16 | 6.09 | 6.04 | 6.00 | 5.96 | 5.91 | 5.86 | 5.80 | 5.77 | 5.75 | 5.72 | 5.69 | 5.66 | 5.63 |
| 5 | 6.61 | 5.79 | 5.41 | 5.19 | 5.05 | 4.95 | 4.88 | 4.82 | 4.77 | 4.74 | 4.68 | 4.62 | 4.56 | 4.53 | 4.50 | 4.46 | 4.43 | 4.40 | 4.36 |
| 6 | 5.99 | 5.14 | 4.76 | 4.53 | 4.39 | 4.28 | 4.21 | 4.15 | 4.10 | 4.06 | 4.00 | 3.94 | 3.87 | 3.84 | 3.81 | 3.77 | 3.74 | 3.70 | 3.67 |
| 7 | 5.59 | 4.74 | 4.35 | 4.12 | 3.97 | 3.87 | 3.79 | 3.73 | 3.68 | 3.64 | 3.57 | 3.51 | 3.44 | 3.41 | 3.38 | 3.34 | 3.30 | 3.27 | 3.23 |
| 8 | 5.32 | 4.46 | 4.07 | 3.84 | 3.69 | 3.58 | 3.50 | 3.44 | 3.39 | 3.35 | 3.28 | 3.22 | 3.15 | 3.12 | 3.08 | 3.04 | 3.01 | 2.97 | 2.93 |
| 9 | 5.12 | 4.26 | 3.86 | 3.63 | 3.48 | 3.37 | 3.29 | 3.23 | 3.18 | 3.14 | 3.07 | 3.01 | 2.94 | 2.90 | 2.86 | 2.83 | 2.79 | 2.75 | 2.71 |
| 10 | 4.96 | 4.10 | 3.71 | 3.48 | 3.33 | 3.22 | 3.14 | 3.07 | 3.02 | 2.98 | 2.91 | 2.85 | 2.77 | 2.74 | 2.70 | 2.66 | 2.62 | 2.58 | 2.54 |
| 12 | 4.75 | 3.89 | 3.49 | 3.26 | 3.11 | 3.00 | 2.91 | 2.85 | 2.80 | 2.75 | 2.69 | 2.62 | 2.54 | 2.51 | 2.47 | 2.43 | 2.38 | 2.34 | 2.30 |
| 15 | 4.54 | 3.68 | 3.29 | 3.06 | 2.90 | 2.79 | 2.71 | 2.64 | 2.59 | 2.54 | 2.48 | 2.40 | 2.33 | 2.29 | 2.25 | 2.20 | 2.16 | 2.11 | 2.07 |
| 20 | 4.35 | 3.49 | 3.10 | 2.87 | 2.71 | 2.60 | 2.51 | 2.45 | 2.39 | 2.35 | 2.28 | 2.20 | 2.12 | 2.08 | 2.04 | 1.99 | 1.95 | 1.90 | 1.84 |
| 24 | 4.26 | 3.40 | 3.01 | 2.78 | 2.62 | 2.51 | 2.42 | 2.36 | 2.30 | 2.25 | 2.18 | 2.11 | 2.03 | 1.98 | 1.94 | 1.89 | 1.84 | 1.79 | 1.73 |
| 30 | 4.17 | 3.32 | 2.92 | 2.69 | 2.53 | 2.42 | 2.33 | 2.27 | 2.21 | 2.16 | 2.09 | 2.01 | 1.93 | 1.89 | 1.84 | 1.79 | 1.74 | 1.68 | 1.62 |
| 40 | 4.08 | .23 | 2.84 | 2.61 | 2.45 | 2.34 | 2.25 | 2.18 | 2.12 | 2.08 | 2.00 | 1.92 | 1.84 | 1.79 | 1.74 | 1.69 | 1.64 | 1.58 | 1.51 |
| 60 | 4.00 | 3.15 | 2.76 | 2.53 | 2.37 | 2.25 | 2.17 | 2.10 | 2.04 | 1.99 | 1.92 | 1.84 | 1.75 | 1.70 | 1.65 | 1.59 | 1.53 | 1.47 | 1.39 |
| 120 | 3.92 | 3.07 | 2.68 | 2.45 | 2.29 | 2.17 | 2.09 | 2.02 | 1.96 | 1.91 | 1.83 | 1.75 | 1.66 | 1.61 | 1.55 | 1.50 | 1.43 | 1.35 | 1.25 |
| ∞ | 3.84 | 3.00 | 2.60 | 2.37 | 2.21 | 2.10 | 2.01 | 1.94 | 1.88 | 1.83 | 1.75 | 1.67 | 1.57 | 1.52 | 1.46 | 1.39 | 1.32 | 1.22 | 1.00 |

*(Reproduced from Snedecor and Cochran, 1967)*

191

**Annex 4b.** Variance ratio F. Probability = 0.01

| | 1 | 2 | 3 | 4 | 5 | 6 | 7 | 8 | 9 | 10 | 12 | 15 | 20 | 24 | 30 | 40 | 60 | 120 | ∞ |
|---|---|---|---|---|---|---|---|---|---|---|---|---|---|---|---|---|---|---|---|
| 1 | 4052 | 4999.5 | 5403 | 5625 | 5764 | 5859 | 5928 | 5982 | 6022 | 6056 | 6106 | 6157 | 6209 | 62.35 | 6261 | 6287 | 6313 | 6339 | 6366 |
| 2 | 98.50 | 99.00 | 99.17 | 99.25 | 99.30 | 99.33 | 99.36 | 99.37 | 99.39 | 99.40 | 99.42 | 99.43 | 99.45 | 99.46 | 99.47 | 99.47 | 99.48 | 99.49 | 99.50 |
| 3 | 34.12 | 30.82 | 29.46 | 28.71 | 28.24 | 27.91 | 27.67 | 27.49 | 27.35 | 27.23 | 27.05 | 26.87 | 26.69 | 26.60 | 26.50 | 26.41 | 26.32 | 26.22 | 26.13 |
| 4 | 1.20 | 18.00 | 16.69 | 15.98 | 15.52 | 15.21 | 14.98 | 14.80 | 14.66 | 14.55 | 14.37 | 14.20 | 14.02 | 13.93 | 13.84 | 13.75 | 13.65 | 13.56 | 13.46 |
| 5 | 16.26 | 13.27 | 12.06 | 11.39 | 10.97 | 10.67 | 1046 | 10.29 | 10.16 | 10.05 | 9.89 | 9.72 | 9.55 | 9.47 | 9.38 | 9.29 | 9.20 | 9.11 | 9.02 |
| 6 | 13.75 | 10.92 | 9.78 | 9.15 | 8.75 | 8.47 | 8.26 | 8.10 | 7.98 | 7.87 | 7.72 | 7.56 | 7.40 | 7.31 | 7.23 | 7.14 | 7.06 | 6.97 | 6.88 |
| 7 | 12.25 | 9.55 | 8.45 | 7.85 | 7.46 | 7.19 | 6.99 | 6.84 | 6.72 | 6.62 | 6.47 | 6.31 | 6.16 | 6.07 | 5.99 | 5.91 | 5.82 | 5.74 | 5.65 |
| 8 | 11.26 | 8.65 | 7.59 | 7.01 | 6.63 | 6.37 | 6.18 | 6.03 | 5.91 | 5.81 | 5.67 | 5.52 | 5.36 | 5.28 | 5.20 | 5.12 | 5.03 | 4.95 | 5.86 |
| 9 | 10.56 | 8.02 | 6.99 | 6.42 | 6.06 | 5.80 | 5.61 | 5.47 | 5.35 | 5.26 | 5.11 | 4.96 | 4.81 | 4.73 | 4.65 | 4.57 | 4.48 | 4.40 | 4.31 |
| 10 | 10.04 | 7.56 | 6.55 | 5.99 | 5.64 | 5.39 | 5.20 | 5.06 | 4.94 | 4.85 | 4.71 | 4.56 | 4.41 | 4.33 | 4.25 | 4.17 | 4.08 | 4.00 | 3.91 |
| 12 | 9.33 | 6.93 | 5.95 | 5.41 | 5.06 | 4.82 | 4064 | 4.50 | 4.39 | 4.30 | 4.16 | 4.01 | 3.86 | 3.78 | 3.70 | 3.62 | 3.54 | 3.45 | 3.36 |
| 15 | 8.68 | 6.36 | 5.42 | 4.89 | 4.56 | 4.32 | 4.14 | 4.00 | 3.89 | 3.80 | 3.67 | 3.52 | 3.37 | 3.29 | 3.21 | 3.13 | 3.05 | 2.96 | 2.87 |
| 20 | 8.10 | 5.85 | 4.94 | 4.43 | 4.10 | 3.87 | 3.70 | 3.56 | 3.46 | 3.37 | 3.23 | 3.09 | 2.94 | 2.86 | 2.78 | 2.69 | 2.61 | 2.52 | 2.42 |
| 24 | 7.82 | 5.61 | 4.72 | 4.22 | 3.90 | 3.67 | 3.50 | 3.36 | 3.26 | 3.17 | 3.03 | 2.89 | 2.74 | 2.66 | 2.58 | 2.49 | 2.40 | 2.31 | 2.21 |
| 30 | 7.56 | 5.39 | 4.51 | 4.02 | 3.70 | 3.47 | 3.30 | 3.17 | 3.07 | 2.98 | 2.84 | 2.70 | 2.55 | 2.47 | 2.39 | 2.30 | 2.21 | 2.11 | 2.01 |
| 40 | 7.31 | 5.18 | 4.31 | 3.83 | 3.51 | 3.29 | 3.12 | 2.99 | 2.89 | 2.80 | 2.66 | 2.52 | 2.37 | 2.29 | 2.20 | 2.11 | 2.02 | 1.92 | 1.80 |
| 60 | 7.08 | 4.98 | 4.13 | 3.65 | 3.34 | 3.12 | 2.95 | 2.82 | 2.72 | 2.63 | 2.50 | 2.35 | 2.20 | 2.12 | 2.03 | 1.4 | 1.84 | 1.73 | 1.60 |
| 120 | 6.85 | 4.79 | 3.95 | 3.48 | 3.17 | 2.96 | 2.79 | 2.66 | 2.56 | 2.47 | 2.34 | 2.19 | 2.03 | 1.95 | 1.86 | 1.76 | 1.66 | 1.53 | 1.38 |
| ∞ | 6.63 | 4.61 | 3.78 | 3.32 | 3.02 | 2.80 | 2.64 | 2.51 | 2.41 | 2.32 | 2.18 | 2.04 | 1.88 | 1.79 | 1.70 | 1.59 | 1.47 | 1.32 | 1.00 |

*(Reproduced from Snedecor and Cochran, 1967)*

# Manual of toolbox

## Tool 1. Estimation of total AI in a given situation

### How to use

In Sheet Tool 1, values in the cells C4 to C7 can be changed. The result (Number of AI per year) is calculated in cell C 9.

### Details

Number of AI necessary can be estimated from the following parameters:

Total number of adult female cattle in the area considered,
Estimated proportion of the adult females brought for AI programme,
Average number of inseminations needed for the birth of a calf, and
Average inter-calving period in months.

Formula to calculate the estimated number of AI performed in a year is:

$$\text{Number of AI per year} = \frac{\text{BFP} * \text{COV} * \text{AIC} * 12}{\text{CI}}$$

where, BFP = total number of adult female population
COV = proportion of adult females estimated to be covered by the programme
AIC = average number of AI per calf born
CI = average calving interval in months

## Tool 2. Herd strength of the bull mother farm of the AI organisation

### How to use

In Sheet Tool 2, all values in column B can be changed, except for B6 and B12.

Intermediate result is available in cell B6.

The result **(Bull mothers to be maintained)** is calculated in cell B12.

### Details

The size of the bull mother farm maintained by the AI organisation will depend up on:

- Total number of AI to be carried out annually
- Average number of doses of semen produced per bull
- Average productive life span of the bulls
- Number of bulls procured from outside sources

- Selection intensity applied on male calves (including mortality and selection for growth, production and reproductive parameters)
- Calving rate of the herd

Estimated number of bull mothers in the AI organization's farm is calculated as:

$$\frac{AIY * (1 + WSS)}{DPB * PLB} - BOS * (20000 / (SIM*CRC))$$

where AIY = number of AI to be done annually

WSS = wastage & minimum surplus of semen doses **(in proportion)**

DPB = average number of semen doses produced / bull / year

PLB = average productive life of bulls (years)

BOS = number of bulls procured from out side

SIM = selection intensity on male calves with in the farm (%)

CRC = calving rate of bull mothers (%)

# Tool 3. Selection intensity for milk yield after first calving in a bull mother farm

### How to use

In Sheet Tool 3, all values in cells C5 to C10 can be changed, except for C9. Values permitted in cell C10 are between 2 and 20.

The results are available in cells C13 to C18.

### Details

**HNOS** Number of heifers in the farm. The exercise is based on the number of heifers in the farm.

**RES** Reproductive success. The **proportion** of animals that become conceived during each reproductive cycle. The remaining proportion of animals are culled and removed from the farm

**SUR** Survival rate. The **proportion** of animals those are alive out of the total number of female calves born

**ICP** Average inter calving period (months) of the farm.

**n** Average number of lactation for which the cows are kept in the farm.

The genetic selection is calculated as follows;

$$GS = \frac{HNOS - (0.5 * HNOS * RES * SUR)}{0.5 * HNOS [(SUR * RES)^2 \, CIN + (SUR * RES)^3 \, CIN^2 + ... + (SUR * RES)^n \, CIN^{(n-1)}]}$$

where:

GS = the genetic selection after first lactation (in proportion)
HNOS = defined above
SUR = defined above
RES = defined above
CIN = calving ratio - the ratio between 12 and the average inter calving period (months) of the farm.
n = defined above.

The following results are obtained:

**Genetic selection:** The extent of genetic selection that is possible after the completion of the first lactation with out change in the herd size. When the calculated figure is one and above it indicates that there is no chance for genetic selection.

**Number of heifers:** Total number of heifers with out considering mortality.

**Final heifer number:** Number of heifers after accounting for the mortality.

**Number of cows:** Number of cows that are available in the farm at any given time.

**Female calves:** Number of female calves that are available in the farm at any given time.

**Total strength:** Total number of female stock available in the farm at any given time. This is the sum of heifers after accounting for mortality, cows and female calves.

# Tool 4. Cost comparison between mobile and stationary AI services

## How to use

In Sheet Tool 4, all values in cells B6 to B18 (stationary) & C6 to C18 (mobile) can be changed except for values in B10, B12, C10 and C12.

Intermediate results are available in cells B10, B12, B19, C10, C12 and C19.

The result **(Cost per AI)** is calculated in cells B20 and C20.

## Details

A comparison between the **mobile** and **stationary** systems of AI service delivery can be done using the tool.

The following parameters must be typed in the appropriate columns for stationary and mobile conditions.

- Salary to the AI technician/year
- Adult females in the area (no.)
- Assumed percentage adult females covered by AI
- Average calving interval (m)
- Number AI required per calf born
- Propulsion charge for mobility of the AI technician
- Cost of a dose of semen
- Liquid nitrogen (lit) required per AI unit/year
- Cost of liquid nitrogen /lit
- Depreciation on liquid nitrogen refrigerators
- Depreciation on motor bike

## Tool 5. Herd size for a field progeny testing programme

### How to use

In Sheet Tool 5, all values in cells B4 to B10 can be changed.

The results are available in cells B11 to B15.

### Details

The herd size and other requirements of a progeny testing programme are calculated using this tool.

The following details must be typed in.
- Number of young bulls to be replaced for the AI programme annually according to the breeding plan
- Percentage of bulls selected as breeding bulls in the AI programme out of the bulls tested
- Planned number of completed daughter lactation/bull
- Expected loss of daughters from birth to completion of 1st lactation (%)
- Average number of AI per calf born
- Average calving interval (months) of the population
- Average coverage of AI in the progeny test area

From the above the following results are calculated using the formulae given against each,

$$\text{Young bulls tested / year} = \frac{\text{Young bulls to be replaced annually} * 100}{\text{Percentage selected}}$$

Total number of completed first lactation = Young bulls tested / year * Average number of completed lactation/bull

Number of female calves to be born

$$= \frac{\text{Total number of completed first lactation} * 100}{(100 - \text{Expected loss from calf birth to completion of 1st lactation})}$$

Number of test AI done / year = Number of female calves to be born *
Average number of AI per calf born * 2
Size of the test herd =

$$\frac{\text{Number of female calves to be born} * 2 * \text{Average calving interval (m) of the population}}{12 * \text{Average coverage of AI in the progeny test area}}$$

## Tool 6. Time taken to estimate breeding value of a bull

### How to use

In Sheet Tool 6, all values in cells C5 to C8 can be changed.

The result (**Average age of the bulls when the sire proof is available**) is calculated in cell C9. The unit of measurement is months.

### Details

Average age of the bulls when the sire proof is available can be estimated using the following parameters:

Average age of bulls when test doses are produced,
Time required to complete test AI,
Average age of the daughters at first calving and
Time required for data analysis and sire evaluation.

The formula to calculate the age of the bulls when the sire proof is available is:

Average age of the bulls when the sire proof is available = BAS + TAI + DAC + TDA + (290/30.5)+10

where,
  BAS = average age of bulls when the required number of semen doses for test AI are produced
  TAI = time required to complete test AI
  DAC = average age of the daughters at first calving
  TDA = time required for data analysis and sire evaluation

Note:

- The factor 290/30.5 calculates the number of months for gestation.
- 10 indicate the lactation period in months.

# Tool 7. Breeding value estimation based on individual records

**How to use**

In Sheet Tool 7, values in the cells C4 and C5 can be changed. New cows (rows) can be entered if required. The lactation values in columns B to F are also editable.

Intermediate results are available in columns G and H.
The result **(Breeding value)** is calculated in column I.

**Details**

The estimation of breeding value based on individual records has the following steps

1. Decide on the **heritability** $(h^2_1)$ of the trait and enter it in column C4
2. Decide on the **repeatability** (R) of the trait and enter it on column C5
3. Enter the **average standard lactation milk yield of the herd** for the lactations one to five in the columns (B9 to F9) provided for it
4. Enter the cow number in column A and record all the available standard lactation milk yields in the appropriate columns against the cow.
5. Use the same unit of measurement for the herd average and the individual animal's record

$$\text{Heritability of m lactations } h^2_m = \frac{h^2_1 * m}{(1 + (m - 1) * R)}$$

where
  m = number of lactation milk yield values considered for the cow
  $h^2$ = heritability of milk yield
  R = repeatability of milk yield

The estimated breeding value of the cows will be shown in the column named **Breeding value** (BV), which is calculated as,

$$BV = \frac{\Sigma(IP - CP)}{m} * h^2_m$$

where,
  IP = lactation performance of the cow
  CP = average lactation performance of the herd (same lactation)
  m = number of lactation milk yield values considered for the cow
  $h^2_m$ = heritability of m lactations

# Tool 8. Estimation of breeding value based on half-sib records

**How to use**

In Sheet Tool 8, value in the cell C4 can be changed. New half-sib group (rows) can be entered if required. The values in columns B to D are also editable.

The result **(Breeding value)** is calculated in column E.

**Details**

The simplest method for estimation of breeding value of bulls based on half-sib records is described below. It explains the principle of half-sib analysis. For more accurate results techniques like BLUP and least squares analysis shall be employed.

The steps involved in this method of contemporary comparison is:

1. Decide on the heritability $(h^2)$ of the trait and enter it in column C4.
2. Enter the group size (number of half-sibs) whose average yield is recorded in the column B.
3. Enter the average first standard lactation milk yield of the group of daughters on the column C.
4. Enter the average first standard lactation milk yield of the contemporaries on the column D.

It is required that the lactation yields are corrected for influences other than that of the bulls before breeding value estimations are carried out.

The estimated breeding value of the bulls based on the performance of his daughters will be shown in the column **Breeding value** (BV), which is calculated as:

$$BV = (IP - CP) * \tfrac{1}{2}\sqrt{n/(n + k)}$$

where,

IP = average first standard lactation milk yield of the group of half-sibs

CP = average first standard lactation milk yield of contemporaries

n = number of half-sibs (group size)

k = $(4 - h^2) / h^2$